ACTIVITY MANUAL WITH INTEGRATED REVIEW WORKSHEETS

Edited by

DONNA KIRK

The College of St. Scholastica

Integrated Review Worksheets prepared by

DOUG NERING

Ivy Tech Community College of Indiana

USING & UNDERSTANDING MATHEMATICS: A QUANTITATIVE REASONING APPROACH

SEVENTH EDITION

Jeffrey Bennett

University of Colorado at Boulder

William Briggs

University of Colorado at Denver

Pearson

ISBN-13: 978-0-13-477664-4
ISBN-10: 0-13-477664-X

Contents

Activities

*Spreadsheet template for this activity can be found in MyLabMath or
at www.pearsonhighered.com/mathstatsresources

*Spreadsheet template for this activity can be found in MyLabMath or
at www.pearsonhighered.com/mathstatsresources

Chapter 11
Activity for Unit 11C – The Golden Mean 93
Contributed by: Shane Goodwin, Brigham Young University–Idaho

Integrated Review Worksheets

Dice Breaker
Using Dice to Model the Spread of a Disease

Before you begin:
- Each student receives a standard, 6-sided die and a Record Sheet (next page).
- Each student needs a unique two- or three-digit ID number for this activity. The instructor will describe how ID numbers are assigned.

Procedure: The activity proceeds through six stages, each lasting 3 to 5 minutes. The instructor announces the start and stop of each stage. When Stage 1 begins, each student should do the following:

1. Leave your seat and mingle with other students. Quickly introduce yourself to another student by both name and ID number. Then:
   - Write the other student's ID number in the Stage 1 column of your Record Sheet.
   - Each of the two students rolls his/her die, and…
       i. If the sum showing on the two dice is 5 or less, **circle** the other student's ID number on your sheet.
       ii. If the sum is 6 or more, leave the other student's ID uncircled.
   Example: Suppose you meet Student 21. You write "21" in the Stage 1 column of your Record Sheet, then each of you roll your die. If the sum is 5 or less, you circle the "21" on your sheet; if the sum is 6 or more, you leave it uncircled.
2. As soon as you have finished, move on and introduce yourself to another student. Write this student's ID below the ID of the last student you met. Then repeat steps 1 and 2 by rolling the dice and circling the ID if the sum is 5 or less.
3. Continue this process until the instructor calls an end to Stage 1. Then, after a few seconds break, Stage 2 begins, which works the same way except that you now record results in the Stage 2 column. The activity should continue through Stage 6. You should strive for 3 to 6 encounters within each stage, and try to avoid multiple encounters with the same person (to the extent possible, which depends on group size).

Analysis: After all the stages are complete, the instructor writes all the ID numbers on a board sequentially. One ID is selected randomly and circled on the board to identify the carrier of an infectious disease, such as influenza. If you encountered this student, you were exposed to the disease, but it was transmitted *only* if you circled the student's ID.

- The instructor asks all students who have the carrier's ID circled in Stage 1 to raise their hands. All these students are now also infected, so the instructor circles their ID numbers on the board, and records the total number of infected students at the end of Stage 1.
- The class leader then asks all students to look at their encounters in Stage 2, and raise their hands if they have circled *any* of the infected ID numbers that are now identified on the board. These students are now also infected, so circle their IDs on the board and record the total number infected at the end of Stage 2.
- Continue this process through all the stages. Make a graph showing the number of students infected at the end of each stage.

Discussion: It's possible to do a careful mathematical analysis of the results, but for now, just use your powers of logic and critical thinking to discuss their meaning with your classmates. Here are some questions to help get the discussion started:

1. What do the results tell you about how a disease can spread through a population?
2. The transmission of many diseases can be stopped by simple steps such as hand washing, or in some cases through vaccination. How would the spread be different if many people took these steps? Could the disease be completely eradicated?
3. We didn't have to define the carrier as someone with a disease. For example, we could have defined the carrier to be someone who starts spreading a rumor, or a computer virus. What other kinds of human interactions can be modeled by this activity?
4. The graph you made is one simple example of how we can analyze data quantitatively. How does this type of analysis help improve your understanding of the situation being modeled?

Record Sheet

Your Personal ID Number   _____

Instructions:

1. During each stage, record the ID number of each student you meet in the column for that stage.
2. Roll your die with each person you meet; circle the person's ID number (which you've already written down) if the sum of the two dice is 5 or less; if the sum is greater than 5, leave the ID number uncircled.

Stage 1	Stage 2	Stage 3	Stage 4	Stage 5	Stage 6

Developing Problem Solving Skills

Problem Solving Exercises

1. **Read carefully.** Anna had six apples and ate all but four of them. How many apples were left?

2. **Read carefully.** If there are 12 one-cent stamps in a dozen, how many two-cent stamps are there in a dozen?

3. **Read carefully.** The butcher is six foot, four inches tall and wears size 14 shoes. What does he weigh?

4. **Read carefully.** Is it legal to marry your widow's sister? Explain.

5. **Read carefully.** I am the brother of the blind fiddler, but brothers I have none. How can this be?

6. **Not guilty.** A lady did not have her driver's license with her when she failed to stop at a stop sign and then went three blocks down a one-way street the wrong way. A policeman saw her, but he did not stop her. Explain.

7. **Whose egg?** If Mr. Jones rooster laid an egg in Mr. Gomez' yard, who owns the egg? Explain.

8. **Fathers and sons.** "Brothers and sisters I have none, but that man's father is my father's son." Who is that man?

9. **Small town haircuts.** A visitor arrived in a small Nevada town in need of a haircut. He discovered that there were exactly two barbers in town. One was well-groomed with splendidly cut hair, the other was unkempt with an unattractive haircut. Which barber should the visitor patronize? Explain.

10. **Travel times.** A bus traveled from the terminal to the airport at an average speed of 30 mi/hr. and the trip took an hour and 40 min. The bus then traveled from the airport back to the terminal and again averaged 30 mi/hr. However, the return trip required 100 min. Explain.

11. **Banquet counting.** There were 100 basketball and football players at a sports banquet. Given any two athletes, at least one was a basketball player. If at least one athlete was a football player, how many football players were at the banquet?

12. **From the eighth century.** A man arrived at the bank of a river with a goat, a wolf, and a head of cabbage. His boat holds only himself and one of his possessions. Furthermore, the goat and the cabbage cannot be left alone and the wolf and the goat cannot be left alone. What is the minimum number of trips needed for the man to cross the river with his three possessions? Show how it is done.

13. **Rolling quarters.** Two quarters rest next to each other on a table. One coin is held fixed while the second coin is rolled around the edge of the first coin with no slipping. When the moving coin returns to its original position, how many times has it revolved?

14. **Buy and sell.** Kelly bought a horse for $500 and then sold it for $600. She bought it back for $700 and then sold it again for $800. How much did she gain or lose on these transactions?

15. **Mixed fruit.** One of three boxes contains apples, another box contains oranges, and another box contains a mixture of apples and oranges. The boxes are labeled APPLES, ORANGES and APPLES AND ORANGES, but each label is incorrect. Can you select one fruit from only one box and determine the correct labels? Explain.

16. **Mixed apples.** Three kinds of apples are all mixed up in a basket. How many apples must you draw (without looking) from the basket to be sure of getting at least two of one kind?

17. **Socks in the dark.** Suppose you have 40 blue socks and 40 brown socks in a drawer. How many socks must you take from the drawer (without looking) to be sure of getting (i) a pair of the same color, and (ii) a pair with different colors?

18. **Time flies.** Reuben says, "Two days ago I was 20 years old. Later next year I will be 23 years old." Explain how this is possible.

19. **Rising tide.** A rope ladder hanging over the side of a boat has rungs one foot apart. Ten rungs are showing. If the tide rises five feet, how many rungs will be showing?

20. **Chocolate demographics.** Suppose one-half of all people are chocolate eaters and one-half of all people are women.
    (i)   Does it follow that one-fourth of all people are women chocolate eaters?
    (ii)   Does it follow that one-half of all men are chocolate eaters? Explain.

21. **Measuring problem.** How do you measure exactly 2 gallons of water from a well using a 4-gallon jug and a 7-gallon jug? Assume the containers have no markings, so fractions of the full volume cannot be measured.

22. **Chess family.** A woman, her older brother, her son, and her daughter are chess players. The worst player's twin, who is one of the four players, and the best player are of opposite sex. The worst player and the best player have the same age. If this is possible, who is the worst player? 23.

23. **Subway dating.** A Manhattan fellow had a girlfriend in the Bronx and a girlfriend in Brooklyn. He decided which girlfriend to visit by arriving randomly at the train station and taking the first of the Bronx or Brooklyn trains that arrived. The trains to Brooklyn and the Bronx each arrived regularly every 10 minutes. Not long after he began his scheme the man's Bronx girlfriend left him because he rarely visited. Give a (logical) explanation.

24. **Clock chimes.** If a clock takes 5 seconds to strike 5:00 (with 5 equally spaced chimes), how long does it take to strike 10:00 (with 10 equally spaced chimes)?

25. **Optimal grilling time.** A small grill can hold two hamburgers at a time. If it takes five minutes to cook one side of a hamburger, what is the shortest time need to grill both sides of three hamburgers?

26. **Mixed up babies.** One day in the maternity ward, the nametags for four girl babies became mixed up. (i) In how many different ways could two of the babies be tagged

correctly and two of the babies be tagged incorrectly? (ii) In how many different ways could three of the babies be tagged correctly and one baby be tagged incorrectly?

27. **Coin weighing problem.** How do you find the light counterfeit coin among eight coins in two weighings? Assume that a balance scale is used and that all coins in question have identical appearance. A weighing consists of putting a sample of coins on each pan of the balance and observing whether the pans balance or whether one pan weighs more than the other pan.

28. **Another coin weighing problem.** How do you find the heavy counterfeit coin among 12 coins in three weighings? See Exercise 27 for ground rules.

29. **A bet.** Alex says to you, "I'll bet you any amount of money that if I shuffle this deck of cards, there will always be as many red cards in the first half of the deck as there are black cards in the second half of the deck." Should you accept his bet? Explain.

30. **Family counting.** Suppose that each daughter in your family has the same number of brothers as she has sisters, and each son in your family has twice as many sisters as he has brothers. How many sons and daughters are in the family?

31. **The three switches.** Suppose you are in a room that has three light switches, each connected to exactly one lamp in an upstairs room. How can you determine which switch operates each lamp with only trip to the upstairs room?

32. **Scale calibration.** The zero point on a bathroom scale is set incorrectly, but otherwise the scale is accurate. It shows 60 kg when Dan stands on the scale, 50 kg when Sarah stands on the scale, but 105 kg when Dan and Sarah both stand on the scale. Does the scale read too high or too low? Explain.

33. **Counterfeit golf balls.** Each of ten large barrels is filled with golf balls that all look alike. The balls in nine of the barrels weigh one ounce and the balls in one of the barrels weigh two ounces. You have a scale that measures absolute weight in ounces. With only one weighing on this scale, how can you determine which barrel contains the heavy golf balls?

34. **Stealing pennies.** Alice takes one-third of the pennies from a large jar. Then Bret takes one third of the remaining pennies from the jar. Finally, Carla takes one-third of the remaining pennies from the jar, leaving 40 pennies in the jar. How many pennies were in the jar at the start?

35. **Suspicious survey.** A survey shows that of 100 nurses, 75 play at least soccer, 95 play at least softball, and 50 play both soccer and softball. Is this possible?

36. **A race.** Stann placed exactly in the middle among all runners in a race. Dan was slower than Stan, in 10th place, and Van was in 16th place. How many runners were in the race?

37. **Vacation weather.** During a vacation, it rained on 13 days, but when it rained in the morning, the afternoon was sunny, and every rainy afternoon was preceded by a sunny morning. There were 11 sunny mornings and 12 sunny afternoons. How long was the vacation?

38. **Eavesdropping.** Suppose you overhear the following conversation. If you were Paul, could you determine the ages of Paula's children? How?
    - Paul: How old are your three children?
    - Paula: The product of their ages is 36 and the sum of their ages is the same as today's date.
    - Paul: That is not enough information.
    - Paula: The oldest child also has red hair.

39. **A very old puzzler.** Three guests checked in at a hotel and paid $30 for their room. A while later the desk clerk realized that the room should have cost only $25, so she gave the bellboy $5 to return to the three guests. The bellboy realized that $5 couldn't be divided evenly among the three guests so he kept $2 and gave $1 to each of the guests. It seems that the three guests have now each spent $9 for the room (that makes $27) and the bell boy has $2, for a total of $29. Where is the missing dollar?

40. **Book orders.** Five books of five different colors are placed on a shelf. The orange book is between the gray and pink book, and these three books are consecutive. The gold book is not first on the shelf and the pink book is not last. The brown book is separated from the pink book by two books. If the gold book is not next to the brown book, what is the complete order of the five books?

Scale of the Earth/Moon System

Before you begin:
- You will need two balls, with diameters of approximately 3-inches and 1-inch, to represent Earth and the Moon, respectively. (The actual ratio of Earth diameter to Moon diameter is 3.7 to 1.)
- You will need a ruler or tape measure.
- You will need a calculator.

Part A. Use your 3-inch ball to represent Earth and your 1-inch ball to represent the Moon. *Without doing any calculations,* make a guess about how far apart Earth and the Moon are on this scale; hold your Earth and Moon (one in each hand) to show your guess. Have your partner measure the distance you are holding them apart, and fill-in your name and the measurement of you guess below. Then trade off so that your partner also gets to make a guess.

       Student_____ Measurement of guess _____

       Student_____ Measurement of guess _____

Part B. The actual average Earth–Moon distance is about 384,000 kilometers, and Earth's diameter is about 12,800 kilometers. How many "Earth diameters" is the distance from Earth to the Moon? Show your work.

Part C. Based on your answer to Part B, what is the correct scaled distance of the Moon, using the 3-inch ball as Earth?

Part D. Measure the distance you found in Part C. How does it compare to the guesses you and your partner made in Part A? Write a couple sentences describing the comparison and what you have learned from it.

Part E. The Sun's actual diameter is about 1,400,000 kilometers. How many "Earth diameters" is this? Given your 3-inch Earth, how large of a ball would you need to represent the Sun? Give your answer in feet, and show your work.

Part F. The average Earth–Sun distance is about 149,600,000 km. To represent this distance to scale, how far away would you have to place your 3-inch Earth from your Sun? Give your answer in feet, and show your work.

Part G. Write a few sentences summarizing what you've learned from this scaling exercise. Be sure to explain why this scale, although very useful for visualizing Earth and the Moon, is much less useful for visualizing the solar system.

Part H (Bonus). The Hubble Space Telescope orbits Earth at a maximum altitude of about 600 km. Given your 3-inch Earth, how far above the ball's surface would the Hubble Space Telescope be located? Based on your answer, what can you say about the common belief that Hubble is useful because it is "closer to the stars"?

How Many Grains of Sand on All the Beaches in the World?

Before you begin:
- You will need a small volume measuring device, such as a quarter-teaspoon or milliliter measure.
- You will need a small amount of sand.
- You will need a calculator.

Procedure: The goal of this activity is to estimate the total number of grains of sand on all the beaches on Earth. The following steps should help you make this estimate.

1. First, discuss among your group a method that would allow you to estimate the total number of grains of sand on all beaches, using only the sand and measuring device, and data about beaches on Earth. Continue to the next step when you think you understand the process.

2. Briefly discuss how you can use the measuring device and sand to determine the number of grains of sand that would fit in one cubic meter. Write a brief description of the procedure you will follow, describing any necessary calculations.

3. You will probably have decided to divide up the actual counting of sand grains. Carry out the count. If you get bogged down (for example, because you have too many grains to count), discuss in your group a way to make the task move more quickly. When you are finished, record your results below:

    volume of sand measured = _____ number of grains = _____

4. Carry out any necessary unit conversions to convert your values from #3 (above) to units of sand grains per cubic meter. Show your work.

5. According to the online CIA Factbook, the total length of sandy beach on Earth is about 360,000 kilometers. You can use this value along with estimates of the average beach width and depth to find the total volume of sand on all the beaches on Earth. Discuss the average beach width and depth until you have numbers you all agree are reasonable. Record them below:

    average beach width = _____ averaged depth of sand = _____

6. Based on your estimates in (5), what is the total volume of the beaches on Earth? Be sure to convert your final answer to units of *cubic meters*. Show your work.

7. Now put all your work together to estimate the total number of sand grains on all the beaches on Earth. Show your work. Give your final answer in scientific notation.

Analysis/Discussion

1. Discuss the uncertainty in your estimate. For example, do you think it is accurate to within 10%, or to within a factor of 2, or a factor of 10, or…? Explain.
2. Astronomers estimate that there are 100 billion galaxies in the (observable) universe, each with an average of about 100 billion stars. What is the total number of stars in the universe, and how does it compare to the total number of grains of sand on all the beaches on Earth?
3. How does the number of grains of sand on all the beaches on Earth compare to the number of *atoms* in a drop of water? You can either try to find this number of atoms with a web search, or you can calculate it with the following facts that you may recall from high school chemistry: a typical drop of water weighs a little less than 1 gram; water consists of molecules of H_2O, which means each water molecule contains 3 atoms; 1 *mole* of H_2O weighs 18 grams and contains Avogadro's number of molecules, which is about 6×10^{23} molecules.
4. Discuss a few other problems for which similar estimation techniques could help you figure out something that might have seemed "unknowable" before you started.

Taking Control of Your Finances
A Family Budget Action Plan Activity Sheet

Before you begin: Review the key principles in unit 4A of your textbook regarding the importance of taking control of your finances. Also, carefully read the following brief background description of the Davidson family before starting into the project.

- Gary and Michelle Davidson currently have 2 children (ages 9 and 7) and a golden retriever dog. Gary works as a manager at a local manufacturing plant. His current annual salary is $59,400. They have barely started saving for retirement and are not receiving full matching funds from the company for their 401-k. To qualify for the maximum matching funds from Gary's employer, they would need to <u>triple</u> their current 401-k contributions.
- They tend to spend more than they make each month but are not sure how much more. They feel like they have too much credit card debt, and don't have enough money set aside for family savings in the categories of emergency, Christmas, medical, or vacation.
- In reality, the Davidsons feel like they are not sure where the money seems to be going and would like to start a monthly budget action plan. In summary, they need your help!

Download Spreadsheet: (for student)

- Download or ask your instructor for the file "BudgetActionPlan_Template.xls" which provides you a spreadsheet framework to quantify the Davidson's monthly inflow and outflow. The formulas are already set up and you will be able to use the template to help the Davidson's balance their monthly budget, helping them achieve their financial goals.

Procedure: Build a "status quo" budget for the Davidsons and make a hard copy. That is, find out the status of their current net monthly cash flow. Read the detailed information about their finances and enter the prorated monthly amounts into the categories of income, payroll deductions, savings, and expenses. Make sure each amount is an accurate <u>monthly</u> reflection of their financial situation. *Note*: there may be more information than you will use and the items may not necessarily match the order of the spreadsheet. Be sure to proofread afterwards.

1. Each month Gary's payroll deductions include $367.45 for federal income tax, $162.50 for state income tax, and $378.68 for their FICA (Social Security and Medicare) tax. Currently $34.17 is being deducted and deposited into the Gary's 401-k.
2. Probably one of the biggest concerns for the Davidsons right now is getting on top of the credit-card debt. Their current MasterCard balance of $4,730 is being assessed an 18% APR finance charge and they have a minimum payment of $95 monthly. They are hoping to pay off their debt within two years but this would require that they increase their monthly credit card payment to at least $236.14 to meet their goal.

3. The family bought a home 5 years ago for $250,000. Their loan is for 30 years at a fixed rate of 6% APR and their monthly payment is $1,349. To safeguard against theft, fire, and other calamities, they carry homeowner's insurance with an annual premium of $984. They are insuring their home with a different company than their vehicle insurance company and hope to combine both types of policies with the same insurer to save 15% on premiums. The Davidsons pay $675 every six months for county property taxes on their home and lot.

4. They currently pay a monthly cable TV bill of $65.49, a landline phone bill of $32.50, an Internet provider bill of $43.68, and a cell phone bill of $109.45. Their utility bills for electricity, natural gas, and water/sewer/garbage are $39.56, $89.75, and $53.49 respectively.

5. The Davidson's are a two-vehicle family. They own an older mini-van that is paid for, but recently purchased a brand new Ford F-150 4x4. They had to borrow just under $30,000 at a 7% APR, which resulted in a $586 monthly payment. They have estimated that they spend about $360 per year on vehicle maintenance and their vehicle insurance premium is $468 every six months. Lately, they have been averaging about $150 per month on gasoline for the van and truck combined.

6. Because the family rarely plans out their weekly meals or uses coupons $635 is a typical monthly average on groceries. They eat out quite often and spend about $125 at restaurants. Both Gary and Michelle belong to a local health club and spend $48 per month on membership dues, their clothing expenditures average about $90 for the family, entertainment is $65, household supplies about $35, miscellaneous items $70, and their monthly dog food and vet bills for their golden retriever average about $42. They also pay approximately $60 to various charities each month.

7. Finally, the Davidsons are doing their best to protect the family with health and life insurance policies. Currently they pay $197.89 (their monthly share of the company's health insurance package) and $47.82 (their quarterly life insurance premium).

Analysis: Upon completing and printing out the "status quo" budget, use the spreadsheet to modify the quantities so you can get the net monthly cash flow to zero and achieve as many of the goals as possible from the *Before You Begin* section above. You can do this by trimming and consolidating expenses and possibly recommending some realistic ways to increase income.

Help them make a new budget action plan, print it out, and answer the following:

1. What was the original (status quo) net monthly cash flow for the Davidsons?

   _____

2. What percent of their monthly gross income is dedicated to the combination of their mortgage, homeowner's insurance, and county property taxes? _____

3. What percent of their monthly gross income is dedicated to vehicle expenses?

    _____

4. List some of your creative but realistic recommended changes to their budget:

    • _____   • _____

    • _____   • _____

    • _____   • _____

    • _____   • _____

    • _____   • _____

Discussion:
1. How do you feel about the homeowner's total percentage mentioned in #2 of the *Analysis*? What do financial experts recommend as a maximum percentage?
2. How do you feel about the total vehicle percentage in #3 of the Analysis? How would you find the line between affordable and not affordable in terms of a vehicle?
3. Reflect on the personal impressions you had as you put together the Davidson's original budget and how this activity might impact your future financial decisions.

Compound Interest
A Three Methods Approach

Before you begin:
- Record your name on the sign-up sheet (given by your instructor) on any arbitrary row and copy the Line #, Principal, and APR in the Results Summary below.
- Review the compound interest formula from section 4-B of your text.
- Remember that all formulas in an Excel spreadsheet begin with an equality symbol '='.
- Recall that to reference a cell, you can click on it or type its column letter and row number.

Download Spreadsheet: (for student)
- Download or ask your instructor for the file "Compound_Interest_Template.xls" which provides you a spreadsheet framework. This helps you concentrate on the three methods of calculating the lump sum investment without worrying about the formatting details.

Procedure: Using the template you have downloaded and the prototype figure below, construct a compound interest spreadsheet using three different methods (iteration (steps), formula, and Excel function) that will arrive at the very same balance if properly done. Be sure to type in the Givens box the same principal, compound, and APR as the prototype figure. From Quarter #1 (row 8) and thereafter, you will be building formulas that are flexible enough to accommodate other values you type into the Givens box later.

Iteration (Steps) Method (I)
1. Link by a cell reference, the total cell of Quarter #0 (cell E7) to the given principal value in cell D2.
2. Now let principal from Quarter #1 (cell C8) be referenced to cell E7.
3. For the interest in Quarter #1 (cell D8), create a formula by multiplying the principal (cell C8) by the given APR (cell D4) divided by the given Compounds (cell D3). Note: because you will want to always use the same given APR and compound values even after you copy or fill the formulas down the columns, you must use absolute cell referencing for those cell locations (locking them in). This can be done by pressing F4 (for Windows) or Command-T (for Macintosh) while the cursor is in the middle or at the end of cell references D4 and D3.

Quarter	Principal	Interest	Total
	Givens: Principal	1500	
	Compounds	4	
	APR	7.50%	
0			1500.00
1	1500.00	28.13	1528.13
2	1528.13	28.65	1556.78
3	1556.78	29.19	1585.97
4	1585.97	29.74	1615.70
5	1615.70	30.29	1646.00
6	1646.00	30.86	1676.86
7	1676.86	31.44	1708.30
8	1708.30	32.03	1740.33
9	1740.33	32.63	1772.96
10	1772.96	33.24	1806.21
11	1806.21	33.87	1840.07
12	1840.07	34.50	1874.57
13	1874.57	35.15	1909.72
14	1909.72	35.81	1945.53
15	1945.53	36.48	1982.01
16	1982.01	37.16	2019.17
17	2019.17	37.86	2057.03
18	2057.03	38.57	**2095.60**
19	2095.60	39.29	2134.89
20	2134.89	40.03	2174.92
21	2174.92	40.78	2215.70
22	2215.70	41.54	2257.25
23	2257.25	42.32	2299.57
24	2299.57	43.12	2342.69
25	2342.69	43.93	2386.61
26	2386.61	44.75	2431.36
27	2431.36	45.59	2476.95
28	2476.95	46.44	2523.39
29	2523.39	47.31	2570.71
30	2570.71	48.20	2618.91
31	2618.91	49.10	2668.01
32	2668.01	50.03	2718.04
33	2718.04	50.96	2769.00
34	2769.00	51.92	2820.92
35	2820.92	52.89	2873.81
36	2873.81	53.88	2927.69
	Total:	$ 1,427.69	

Methods:
- Qtr. #18 *Iteration (Steps)* "=C25+D25"
- Qtr. #18 *Formula* **2095.60** "=1500*(1+0.075/4)^18"
- Qtr. #18 *Excel Function* **2095.60** "-FV(D4/D3,B25,0,D2)"

4. Now add the interest to the principal (that is, compound the interest) to make a new total (cell E8). This is done by adding cell C8 to cell D8 as a formula (that is, "=C8+D8").

5. Complete the Iteration Method by dragging over the principal, interest, and total cells from Quarter #1 and filling their corresponding formulas down to Quarter #36. This can be done relatively fast by looking for the small black cross hair in the bottom right corner of your shaded rectangle and either dragging the three cells down or simply double-clicking. See your software documentation or instructor for help. Don't forget to check for accuracy.

Formula Method (II)
1. In the Formula box at the bottom of your template, follow the directions for the compound interest formula as shown in the prototype.
2. Confirm that your answer matches the same one found for Quarter #18 using iterations.

Excel Function Method (III)
1. In the Excel Function box at the bottom of your template, follow the directions for the formula as shown in the prototype. The focus here is to learn how to use the built-in function known as future value (FV) for Excel. You can find helpful dialog boxes to guide you through this process by looking for the "Insert" ribbon in the newer Office suite for Windows or by using the "Insert" pull-down menu and finding the feature "Function…" in the Macintosh Office suite.
2. You will notice that the template and prototype indicate cell references in the future value function rather than actual values. This reinforces the idea of keeping your formula generalized to accommodate future changes to the Givens box.
3. Be aware of the need to place a negative sign after the equality to keep the future value as a positive value.

Polishing the Worksheet
1. Be sure to total your interest after Quarter #36 (D44) by using the function 'Sum' and dragging over the interest cells from Quarter #1 to Quarter #36 or take advantage of the Auto Sum tool that looks like Σ and may be on one of your tool bars at the top.
2. Check to see if the very last balance (Quarter #36) matches the prototype quantity.

Results Summary: Upon verifying that your compound interest spreadsheet (based on the prototype givens) has correct quantities throughout, modify your spreadsheet according to the sign-up sheet and/or directions provided by your instructor and complete the following:

Your Sign-up Numbers:

Line # _____   Initial Balance (Principal): _____   APR: _____

Quarter #18 Total:

Method I: _____   Method II: _____   Method III: _____

Quarter #36 Total: _____   Interest Total: _____

Discussion:

1. What would be another way to calculate total interest rather than just adding up the individual interest amounts?
2. What percentage of the balance (end of Quarter #36) is the interest and what percentage is the principal? Do these percentages change if you modify the principal or APR? How?
3. What are some of the pros and cons to each of the three methods you have used?
4. Which of the three methods is your favorite way to calculate the balance and why?
5. What impressions or insights have you gained from this exercise?

Regular Payments Savings Plan
A Three Methods Approach

Before you begin:
- Record your name on the sign-up sheet (given by your instructor) on any arbitrary row and copy the Line #, Principal, APR, and Payment in the Results Summary below.
- Review the savings plan formula (regular payments) from section 4-C of your text.
- Remember that all formulas in an Excel spreadsheet begin with an equality symbol '='.
- Recall that to reference a cell, you can click on it or type its column letter and row number.

Download Spreadsheet:
- Download or ask your instructor for the file "Savings_Plan_Template.xls" which provides you a spreadsheet framework. This helps you concentrate on the three methods of calculating the lump sum investment without worrying about the formatting details.

Procedure: Using the template you have downloaded and the prototype figure below, construct a savings plan spreadsheet using three different methods (iteration (steps), formula, and Excel function) that will arrive at the very same balance if properly done. Be sure to type in the Givens box the same principal, compound, and APR as the prototype figure. From Month #18 (row 26) and thereafter, you will be building formulas that are flexible enough to accommodate other values you type into the Givens box later.

Iterations (Steps) Method

1. Link by a cell reference, the total cell of Month #0 (cell F8) to the principal in the Givens box (cell D2).
2. Now let principal from Month #1 (cell C9) reference from the total (cell F8).
3. For the interest in Month #1 (cell D9), create a formula by multiplying the principal (cell C9) by the given APR (cell D4) divided by the given Compounds (cell D5). Note: because you will want to always use the same given APR and compound values even after you copy or fill the formulas down the columns, you must use absolute cell

Givens:

Principal	500.00
Payment	25.00
APR	4.50%
Compounds	12

Month	Principal	Interest	Payment	Total
0				500.00
1	500.00	1.88	25.00	526.88
2	526.88	1.98	25.00	553.85
3	553.85	2.08	25.00	580.93
4	580.93	2.18	25.00	608.11
5	608.11	2.28	25.00	635.39
6	635.39	2.38	25.00	662.77
7	662.77	2.49	25.00	690.25
8	690.25	2.59	25.00	717.84
9	717.84	2.69	25.00	745.54
10	745.54	2.80	25.00	773.33
11	773.33	2.90	25.00	801.23
12	801.23	3.00	25.00	829.24
13	829.24	3.11	25.00	857.35
14	857.35	3.22	25.00	885.56
15	885.56	3.32	25.00	913.88
16	913.88	3.43	25.00	942.31
17	942.31	3.53	25.00	970.84
18	970.84	3.64	25.00	999.48
19	999.48	3.75	25.00	1028.23
20	1028.23	3.86	25.00	1057.09
21	1057.09	3.96	25.00	1086.05
22	1086.05	4.07	25.00	1115.12
23	1115.12	4.18	25.00	1144.30
24	1144.30	4.29	25.00	1173.60
25	1173.60	4.40	25.00	1203.00
26	1203.00	4.51	25.00	1232.51
27	1232.51	4.62	25.00	1262.13
28	1262.13	4.73	25.00	1291.86
29	1291.86	4.84	25.00	1321.71
30	1321.71	4.96	25.00	1351.66
31	1351.66	5.07	25.00	1381.73
32	1381.73	5.18	25.00	1411.91
33	1411.91	5.29	25.00	1442.21
34	1442.21	5.41	25.00	1472.62
35	1472.62	5.52	25.00	1503.14
36	1503.14	5.64	25.00	1533.78
Total:		133.78	900.00	

Methods

Month #18	Iterations (Steps)	999.48
	'=C24+D24+E24	
Month #18	Formula (Regular Payments)	999.48
	'=500*(1+0.045/12)^18+25*((1+0.045/12)^18-1)/(0.045/12)	
Month #18	Excel Function	999.48
	'=FV(D5/D4,B26,D3,D2)	

referencing for those cell locations (locking them in). This can be done by pressing F4 (for Windows) or Command-T (for Macintosh) while the cursor is in the middle or at the end of cell references D4 and D5.

4. Now add the interest to the principal (that is, compound the interest) along with the payment from Month #1 (cell E9) to make a new total balance (cell F9). This is done by adding cells C9, D9, and E9 as a formula (that is, '=C9+D9+E9' or '=sum(C9:E9)').

5. Complete the Iteration Method by dragging over the principal, interest, payment, and total cells from Month #1 and filling their corresponding formulas down to Month #36. This can also be done by looking for the small black cross hair in the bottom right corner of your shaded rectangle and dragging the three cells down or simply double-clicking. See your software help or instructor for help. Don't forget to check for accuracy.

Formula Method
1. In the Formula box at the bottom of your template, follow the directions for the Savings Plan (Regular Payments) formula as shown in the prototype.
2. Confirm that your answer matches the same one found for Month #18 using iterations.

Excel Function Method
1. In the Excel Function box at the bottom of your template, follow the directions for the formula as shown in the prototype. The focus here is to learn how to use the built-in function known as future value (FV) for Excel. You can find helpful dialog boxes to guide you through this process by looking for the "Insert" ribbon in the newer Office suite for Windows or by using the "Insert" pull-down menu and finding the feature "Function…" in the Macintosh Office suite.
2. You will notice that the template and prototype indicate cell references in the future value function rather than actual values. This reinforces the idea of keeping your formula generalized to accommodate future changes to the Givens box.
3. Be aware of the need to place a negative sign after the equality to keep the future value as a positive value.

Polishing the Worksheet
Be sure to total your interest after Month #36 (D45) and payments (E45) by using the function 'Sum' and dragging over the interest cells from Month #1 to Month #36 or take advantage of the Auto Sum tool that may be on one of your tool bars. Check to see if the very last balance (Month #36) matches the prototype quantity.

Results Summary: Upon verifying that your savings plan spreadsheet (based on the prototype givens) has correct quantities throughout, modify your spreadsheet according to the sign-up sheet and/or directions provided by your instructor and complete the following:

<u>Your Sign-up Numbers:</u>

Line # _____   Initial Balance (Principal): _____   APR: _____   Payment: _____

<u>Month #18 (3 Methods):</u>

Iterations: _____   Formula: _____   Excel Built-in: _____

Month #36 Balance:_____   Interest Total: _____   Payment Total: _____

Discussion:

1. What would be another way to calculate total interest rather than merely using '=sum('?
2. When was the original principal invested and when was the payment invested (end or beginning of the period)? In the finance world, what type of investment is this called?
3. What are some of the pros and cons to each of the three methods you have used?
4. Which of the three methods is your favorite way to calculate the balance and why?
5. What impressions or insights have you gained from this exercise?

Home Loan Amortization, Part I
The Tug-of-War between Interest and Principal

Before you begin:
- Do an Internet search for a website such as www.bankrate.com or www.quickenloans.com where you can investigate the current fixed rates for a home mortgage.
- Do an Internet search for a real estate website at a location of interest to you and find a home that is priced close to the median price for residential homes in that area. Note: you can get a sense of median prices throughout the nation by looking at websites such as www.housingtracker.com or www.kiplinger.com/tools/houseprices/.

Download Spreadsheet: (for student)
- Download or ask your instructor for the file "Amortization_Template.xls" providing you a spreadsheet framework. This helps you concentrate on the flow of the principal and interest payments each month without worrying about the formatting details.

Procedure: Using the template you downloaded and the given quantities typed from the prototype figure below, construct an installment loan (amortization) from scratch as follows:

Calculations Box

1. To calculate the Down Payment $ (cell D6), multiply the Down Payment % (cell D3) by the house sale price (cell D2).
2. Find the Loan Amount (cell D7) by subtracting the Down Payment $ (cell D6) from the House Sale Price (cell D2).
3. To properly create the Payment (cell D8) you will want to use the built-in function '=pmt'. Note: You can find helpful dialog boxes to guide you through this process by looking for the "Insert" ribbon in the Office suite for Windows or, for the Macintosh Office suite, by using the "Insert" pull-down menu and finding the feature "Function…". To avoid the negative sign in the body of the loan schedule, place a negative sign between the equality sign and the Excel function 'pmt'.

	A	B	C	D	E
1					
2		Givens	House Sale Price	235,000.00	
3			Down Payment %	10.00%	
4			APR	6.00%	
5			Loan Years	30	
6		Calculations	Down Payment $	23,500.00	
7			Loan Amount	211,500.00	
8			Payment	1,268.05	
9			Total Interest	244,997.77	
10					
11					
12		Month	Interest	Principal	Balance
13		0			211500.00
14		1	1057.50	210.55	211289.45
15		2	1056.45	211.60	211077.85
16		3	1055.39	212.66	210865.19
17		4	1054.33	213.72	210651.46
18		5	1053.26	214.79	210436.67
19		6	1052.18	215.87	210220.81
20		7	1051.10	216.95	210003.86
21		8	1050.02	218.03	209785.83
22		9	1048.93	219.12	209566.71
23		10	1047.83	220.22	209346.50
24		11	1046.73	221.32	209125.18
25		12	1045.63	222.42	208902.76
26		13	1044.51	223.54	208679.22
27		14	1043.40	224.65	208454.57
28		15	1042.27	225.78	208228.79
29		16	1041.14	226.91	208001.88
30		17	1040.01	228.04	207773.84
31		18	1038.87	229.18	207544.66
32		19	1037.72	230.33	207314.34
33		20	1036.57	231.48	207082.86
34		21	1035.41	232.64	206850.23
35		22	1034.25	233.80	206616.43
36		23	1033.08	234.97	206381.46
37		24	1031.91	236.14	206145.32

4. To calculate the Total Interest (cell D9) you can choose between a built-in function approach using '=cumipmt' or perhaps the easier method of simply multiplying the Payment (cell D8) by the Loan Years (cell D5) and by 12 and then subtracting the Loan Amount (cell D7). This, in essence, is simply taking the total payback amount and pulling out the principal. What is left over represents the amount of cumulative

interest paid over the many months of the loan.

Body of the Amortization Schedule

1. Link by a cell reference, the 0th month's balance (cell E13) to the Loan Amount in the Calculations box (cell D7). This will accommodate any changes in the loan setup boxes that you might make later on.
2. In the tug of war between interest and principal, we always calculate the amount of interest for the month first and then go on to calculate the principal. For the amount of interest to pay in Month #1 (cell C14), create a formula by multiplying the balance from the previous month (cell E13) by the given APR (cell D4) divided by 12 (since most amortization schedules will be based on months). Note: because you will want to always use the same given APR as you copy or fill the formulas down the columns, you must use absolute cell referencing on cell D4. This can be done, by pressing F4 (for Windows) or Command-T (for Macintosh) while the cursor is in the middle or at the end the text D4. Resulting dollar signs indicate to Excel that the particular specific cell will always be referenced rather than allowing a shift of cells as the formula is replicated.
3. Now calculate the principal portion of the payment by subtracting the Interest (cell C14) from the Payment (cell D8) using an absolute cell reference on D8 since it too will always remain the payment throughout the loan (see comments above).
4. For Balance (cell E14), subtract Principal (cell D14) from previous Balance (cell E13).
5. To complete the rest of the months, select the interest, principal, and balance cells from Month #1 (cells C14, D14, and E14) and fill their corresponding formulas down to the last month of your loan. This can also be done by looking for the small black cross hair in the bottom right corner of your shaded rectangle and dragging the three cells down or simply double-clicking. See your software documentation or instructor for help.

Analysis: Upon verifying the accuracy of your loan schedule (based on the prototype givens), modify your spreadsheet according to the house you have chosen and a realistic APR that you have investigated. Now answer the following using your completed spreadsheet:

1. Which is the approximate month in which your balance finally reached half of the original loan amount? Why is it not simply the halfway point in the length of the loan?

2. Calculate the amount you could save in interest by financing over 15 years instead of 30.

3. Calculate interest saved if the APR were 1 percentage point lower than the one you used.

4. Calculate interest saved if a 20% down payment were made on the home instead of 10%.

Discussion:
1. What impresses you about the shape of the balance curve found on the scatterplot worksheet? That is, notice the speed at the loan repayment and how it changes.

2. When in the loan is the majority of the interest paid back to the lending institution?
3. If so much interest can be saved by apply for a 15-year mortgage as opposed to a 30-year mortgage, why do more people not do it?
4. Considering that when you get a loan from a lending institution they probably provide you a copy of the actual amortization, describe why there is still value in knowing how to build an amortization schedule from scratch as you have done in this project.

Home Loan Amortization, Part II
Extra Payment Scenario

Before you begin:
- Complete the Home Loan Amortization, Part I activity in order to understand the role that interest and principal play in the calculations of each month's payment.
- Have your previously completed Amortization_Template.xls ready to expand as you go through this activity.

Procedure: Using the template you completed in the previous activity and the given quantities from the prototype figure below, expand your installment loan sheet as follows:

Extra Payment Scenario Box
1. Type in the Extra Principal (cell J3) $105.67 from the prototype below. This represents the regular payment divided by 12 (allowing for one full payment to be paid per year). Later you can increase or decrease this extra amount according your personal goals.
2. To calculate the Months to Pay (cell J4)—based on the regular payment plus extra—insert the built-in function '=nper' which requires the following inputs: APR divided by 12 (D4/12); both the Payment and Extra Principal as negatives (–D8 – J3) or the negative sum of the payments would also work –(D8 + J3); present value, that is, the Loan Amount (D7). The rest of the inputs are not needed since the defaults are properly set such as the zero future value suggesting that the debt will be paid in full.
3. In order to properly count the Months Saved (cell J5), simply multiply the number of years of the original loan (cell D5) by 12 and subtract the Months to Pay (cell J4).
4. Years Saved (cell J6) is Months Saved (cell J5) divided by 12.
5. New Interest (cell J7) is calculated by multiplying the quantity of Payment and Extra Principal (cells D8 + J3) by Months to Pay (cell J4) and subtracting Loan Amount (cell D7).
6. For Interest Saved (cell J8), simply subtract New Interest (cell J7) from Total Interest (cell D9).

	A	B	C	D	E	F	G	H	I	J	K
1											
2			House Sale Price	235,000.00					Extra Payment Scenario:		
3		Givens	Down Payment %	10.00%					Extra Principal	105.67	
4			APR	6.00%					Months to Pay	295	
5			Loan Years	30					Months Saved	65	
6			Down Payment $	23,500.00					Years Saved	5.5	
7		Calculations	Loan Amount	211,500.00					New Interest	193,062.83	
8			Payment	1,268.05					Interest Saved	51,934.94	
9			Total Interest	244,997.77							
10											
11											New
12		Month	Interest	Principal	Balance		Month	Interest	Principal	Extra Principal	Balance
13		0			211500.00		0				211,500.00
14		1	1057.50	210.55	211289.45		1	1,057.50	210.55	105.67	211,183.78
15		2	1056.45	211.60	211077.85		2	1,055.92	212.13	105.67	210,865.98
16		3	1055.39	212.66	210865.19		3	1,054.33	213.72	105.67	210,546.59
17		4	1054.33	213.72	210651.46		4	1,052.73	215.32	105.67	210,225.60
18		5	1053.26	214.79	210436.67		5	1,051.13	216.92	105.67	209,903.01
19		6	1052.18	215.87	210220.81		6	1,049.52	218.53	105.67	209,578.80
20		7	1051.10	216.95	210003.86		7	1,047.89	220.16	105.67	209,252.98
21		8	1050.02	218.03	209785.83		8	1,046.26	221.78	105.67	208,925.52
22		9	1048.93	219.12	209566.71		9	1,044.63	223.42	105.67	208,596.43
23		10	1047.83	220.22	209346.50		10	1,042.98	225.07	105.67	208,265.69
24		11	1046.73	221.32	209125.18		11	1,041.33	226.72	105.67	207,933.30
25		12	1045.63	222.42	208902.76		12	1,039.67	228.38	105.67	207,599.25

Body of the Amortization Schedule

1. Link by a cell reference, the 0th month's balance (cell K13) to the loan amount in the Calculations box (cell D7).

2. As was the case in activity in Part I, you must calculate the Interest first for Month #1 (cell H14). Multiply the balance from the previous month (cell K13) by the given APR (cell D4) divided by 12 and don't forget to absolute cell reference D4 since it must be used all the way down the amortization.

3. Now calculate the Principal portion of the payment (cell I14), by subtracting the interest (cell H14) from the monthly payment (cell D8) using an absolute cell reference on D8 for the same reason as stated above regarding the APR.

4. For the Extra Principal (cell J14), make an absolute cell reference back to Extra Principal (cell J3).

5. The New Balance (cell K14) can be calculated by subtracting the sum of Principal and Extra Principal (cells I14 + J14) from the previous month's New Balance (cell K13).

6. To complete the rest of the months, select the Interest, Principal, Extra Principal, and New Balance cells from Month #1 (cells H14, I14, J14, and K14) and fill their corresponding formulas down to the last month of your loan—corresponding to the number of months indicated in your Extra Payment Scenario Box under Months to Pay (cell J4). You will need to modify the last Principal payment to zero out the balance.

Analysis: Upon verifying the accuracy of your loan schedule (based on the prototype givens), modify your spreadsheet according to the house you have chosen and a realistic APR that you have researched. Answer the following using your completed spreadsheet:

1. How much time in years and months have you saved by paying one extra payment per year?

2. How much interest have you saved by doing this?

3. With trial and error using your Extra Principal box, how much extra would you need to pay to save just two years on your loan?

4. How much interest would this save you and was it surprising?

5. If you doubled your payment each month by paying an Extra Principal equivalent to the original payment, would you pay your loan off in half the time? If not, why not?

Discussion:
1. As you scroll through the amortization comparing the original balance to the new, what impresses you overall?
2. What are some options available to a homeowner, perhaps later on, that might make as much of an impact on the debt as paying extra principal?

A Case Study of Paying Extra Principal on a Mortgage
Great Idea or "Height of Foolishness"?

Before you begin: Review the loan basics in unit 4D of your textbook regarding the payment formula and the roles that interest and principal play in an amortization. Then carefully read the following pieces of advice written by two different nationally syndicated financial columnists (Sharon Epperson and Bruce Williams) regarding paying extra principal on a mortgage:

- **Sharon Epperson** from Money Smart of USAWeekend.com

Pay mortgage early?
Q: My husband and I are 49 and 48 and are paying extra on our mortgage to have it paid off by the time we're 56. My friend says that will hurt us on our taxes; my husband says it's better to save the money in interest now than to worry about a mortgage tax deduction later. What's your take? S.R. Sheboygan, Wis.

A: You married a smart man. Paying off your mortgage early will save thousands of dollars, and you'll get a reliable rate of return on your investment (you save the interest you would have paid on your mortgage). Yes, you'll lose the mortgage interest tax deduction when that happens. But if you're in the 25% tax bracket, for example, you'd only get back a quarter for each $1 in interest you pay --not such a big break. If you're debt-free and maxing out your 401(k) and IRAs, which offer tax breaks, paying up early isn't a bad idea.

Source: http://www.usaweekend.com/08_issues/080511/080511thinksmart-mortgage-broadwaytickets.html (Posted May 11, 2008).

- **Bruce Williams** from Smart Money

DEAR BRUCE: I have a friend who says that you often advise that it is not wise to pay off a mortgage in advance. Could you tell me in one paragraph why this is a bad idea? It seems to me that being debt-free is a goal worth working toward. – L.H. Syracuse, N.Y.

DEAR L.H.: In a nutshell, the cheapest money that you can borrow is against a first mortgage on your primary residence. Generally speaking, it's in a sub-7 percent range today. It is not too difficult to earn substantially more than that in the marketplace, so why pay off the loan early and settle for an effective return of below 7 percent? You could invest this money elsewhere at a far better return. In addition, if you itemize the interest that you are paying on the home loan, it becomes a deductible item. To me, it's a no-brainer. For younger people to pay off a mortgage early is, in my opinion, the height of foolishness. When you get into your 60s and the idea of having your home paid for out distances the need for return, I have no objection.

Source: "Should 35-year-old save for down payment or retirement?" *Post Register*, July 19, 2001.

Download Spreadsheet: (for student)
- Download or ask your instructor for the file "Financial_Toolboxes.xls" which provides you a collection of mini financial calculators and a worksheet on the role of the negative.

Procedure:

In our brief case study, we assume the Thomas and Jefferson families have identical mortgages (30-year term, fixed-rate 6% APR, and a loan amount of $175,000).The Thomas family will not pay extra but the Jeffersons will. Follow the steps below prior to your analysis.

1. Using the Payment mini calculator of the Financial Toolboxes spreadsheet, calculate the mortgage payment (the same for both families).

   Required Monthly Payment = $_____

2. Assume that the Thomas's will make only the required mortgage payment. The Jeffersons, however, would like to pay off their loan early. They decide to make the equivalent of an extra payment each year by adding an extra 1/12 of the payment to the required amount.

   Complete the following calculations to find what they plan to pay each month:

   a. 1/12 of the required monthly payment = $_____

   b. By adding this 1/12 to the required payments, the Jeffersons plan to pay $_____ each month.

3. The Thomas's will take the full 30 years to pay off their loan, since they are making only the required payments. The Jefferson's extra payment amount, on the other hand, will allow them to pay off their loan more rapidly. Use the Years mini financial calculator of the Financial Toolbox spreadsheet to calculate the approximate number of years (nearest 10th) it would take the Jeffersons to pay off their loan.

   Number of years to pay off loan = _____

Analysis: *For the Thomas Family:* assume that they could afford to make the same extra payment as the Jeffersons, but instead they decide to put that money (#2a. from Procedures above) into a savings plan called an annuity. Use the Future Value mini financial calculator of the Financial Toolbox spreadsheet to calculate how much they will have in their savings plan at the end of 30 years at the various interest rates. Write your answers (to the nearest dollar) in the appropriate cells of the table below.

For the Jefferson Family: assume that they save nothing until their loan is paid off, but then after their debt is paid, they start putting their full monthly payment and 1/12 (#2b. from Procedures above) into a savings plan. The time in months they invest is equal to360 months minus the number of months needed to pay off the loan (#3 from Procedures above) multiplied by 12. Use the Future Value mini financial calculator to calculate how much they will have in their savings plan at the various interest rates. Write your answers (to the nearest dollar) in the appropriate cells of the table below.

	Thomas Family			Jefferson Family	
Rates	1/12th of Monthly Payment _____ Annuity Amount in 360 Months		Rates	Monthly Payment + Extra 1/12th _____ Annuity Amount in 360 Months	
0%			0%		
1%			1%		
2%			2%		
3%			3%		
4%			4%		
5%			5%		
6%			6%		
7%			7%		
8%			8%		

Discussion:

1. What generalizations can you make from the annuity amounts reflected in the analysis table above with regards to the different strategies taken by the families? That is, from a purely financial aspect of the calculations in your table what generalizations could you make regarding the two different strategies?

2. What assumptions may not necessarily be valid for a typical family regarding both the loan rate and savings plan rate?

3. Discuss (as directed by your instructor)some basic pros and cons to these two very different approaches the Thomas and Jefferson families made with their extra monthly payment. Consider various ideas such as possible changes in the family's employment situation, market performance, tax deductions, etc.

4. Comment on the merits of the advice you read from the two financial columnists.

5. Note the dates of the advice columns. How might market performance figure in to their advice they gave at that time?

6. Why do you think Sharon Epperson's advice at the end specifically calls attention to an assumption of whether you are "debt-free and maxing out your 401(k) and IRAs?"

7. If you were to pay extra principal on a mortgage, when is the best time to do it (early or later in the loan process) and why?

8. When you pay extra principal on a loan, describe whether you feel you are actually earning interest on that money or not. That is, how does the old adage "a penny saved is a penny earned" apply in this context?

9. [Bonus] Rework your calculations using a different starting interest rate for the mortgage and/or a different extra payment amount. Do these changes affect any of the generalizations you have made above? Explain.

U.S. Federal Income Marginal Tax Rates
An Amazing History of Change

Before you begin: Review the income tax basics in unit 4-E of your textbook regarding the calculation of the federal income tax based on a given taxable income. Familiarize yourself with the 2013tax rates found in Table 4.9 of your text. This activity is organized into three simple parts addressing 1) the relationship of the actual IRS 1040 tax tables to your 4-E homework, 2) the comparison of the 2000 federal income tax rates to those of 2009 at two different income levels, and 3) the historical overview of U.S. marginal tax rates since 1913.

Download Spreadsheet: (for student)
Download or ask your instructor for the file "Marginal_Tax_Rate_Analysis_History.xls" which provides you a spreadsheet of the upper tax brackets for the federal income tax since 1913 (including summary statistics and a scatterplot to demonstrate the evolution of changes).

Procedure and Analysis:
1. For taxable incomes of $100,000 or less, the *1040 Forms & Instructions* booklet displays tax tables with the income tax previously calculated for the applicable income level (up to $100,000) and the filing status. Locate a copy of the tax tables on the Internet with the search keywords "2009 1040 tax tables" or from the Forms and Publications link found at the website www.irs.gov. Using $100,000 as a taxable income, find the income tax for a couple (married and filing jointly) in the published tax tables. Record your results.

   a. Total Income Tax (based on 2009 IRS Tax Tables) = $_____

   b. Now, calculate the income tax on the very same $100,000 taxable income using the computations described in unit 4-E. Use the framework below to aid in your calculations:

Tax Rate	Married Filing Jointly	Tax Computation
10%	up to 16,700	
15%	up to 67,900	
25%	up to 137,050	
	Total Income Tax:	

   c. Should your answers for Parts (a) and (b) match exactly? Explain why or why not?

2. a. Congress has changed tax rates many times, and one example of major change occurred in the summer of 2001, during the Bush (43) administration. The legislation was officially titled *The Economic Growth and Tax Relief Reconciliation Act of 2001*. The tables below contrast the tax rates before this legislation (in 2000) and later, in 2009. Calculate the income tax for both 2000 and 2009 using a taxable income of **$400,000** and married filing jointly status.

Also calculate the overall tax percentage.

2000 Tax Rate	Married Filing Jointly	Tax Computation
15%	up to 43,850	
28%	up to 105,950	
31%	up to 161,450	
36%	up to 288,350	
39.6%	over 288,350	
	Total Income Tax:	
	% of Taxable Income:	

2009 Tax Rate	Married Filing Jointly	Tax Computation
10%	up to 16,700	
15%	up to 67,900	
25%	up to 137,050	
28%	up to 208,850	
33%	up to 372,950	
35%	over 372,950	
	Total Income Tax:	
	% of Taxable Income:	

b. Now, calculate the income tax for both years 2000 and 2009 based on a taxable income of **$50,000** (married filing jointly).

2000 Tax Rate	Married Filing Jointly	Tax Computation
15%	up to 43,850	
28%	up to 105,950	
	Total Income Tax:	
	% of Taxable Income:	

2009 Tax Rate	Married Filing Jointly	Tax Computation
10%	up to 16,700	
15%	up to 67,900	
	Total Income Tax:	
	% of Taxable Income:	

c. Calculate the absolute and relative change between Total Income Tax for the two incomes ($400,000 and $50,000) between the years 2000 and 2009. Also, calculate the percentage point change in the overall % of taxable income.

	Taxable Income	
Between 2000 and 2009	$400,000	$50,000
Absolute change of total income tax (in dollars):		
Relative change of total income (as a percent):		
Percentage point change in the % of taxable income:		

d. Comment on the changes in the two tax brackets (2000 vs. 2009). Why do you think one brackets in one income level changed more than the other? Do you think the changes made the tax system fairer or less fair? Defend your opinion.

e. On the Internet, search for the current federal income tax brackets (use the current year combined with the keywords "1040 tax tables" and scroll all the way through the tax tables past the $100,000 taxable income to find the bracket structure). Comment on any changes that may have taken place since 2009. If they differ in any substantial way, research why Congress made those changes.

3. Using the uploaded spreadsheet "Marginal_Tax_Rate_History.xls", answer the following:

a. In what year(s) was the upper tax bracket the highest and what percentage was it? Considering the year(s), make a conjecture as to why the rate might have been so high.

b. As you look over the scatterplot of the upper tax bracket history, briefly mention what might account for the sudden rise or drops in marginal tax rates for our nation.

c. How does the current year's upper tax bracket (answer from question 2-e) compare with the median rate found in the summary statistics?

d. What general pattern or trend would you say our upper tax bracket has taken since the 1970s and on?

e. What general pattern or trend would you say our lower tax bracket has taken since the 1970s and on?

Discussion:

1. For an even broader overview of our nation's tax bracket history, go to the URL http://en.wikipedia.org/wiki/Income_tax_in_the_United_States#History_of_top_rates.5B 21.5D. Discuss impressions you have as you peruse the historical data, graph, and commentary found in the Wikipedia entry?

2. Investigate the history of tax compliance, that is, did very many people really pay those high rates back in the 1940s and 1950s?

3. Investigate the approximate year in which income taxes were first actually withheld directly from an employee's payroll throughout the year. What were some of the pros and cons of this approach that were debated at the time?

Interpreting Measures of Central Tendency
Data Displays, Frequency Charts and Frequency Graphs

Before you begin:
- Review how to calculate mean, median, and mode.
- Review terms such as frequency.

Background: Everyday, we see results displayed in newspapers, on television, or on the Internet. Understanding how to read graphs and analyzing them is essential in making sense of the world around you. You should be able to organize and analyze data in various displays so as to make sense of the data and to answer questions.

In this activity you will:
1. Calculate mean, median, mode, and range of various data.
2. Explore the measures of center and be able to identify what they are given various data displays (frequency charts and frequency graphs).
3. Apply your knowledge to real-life problems and explain how data can be informative.

Procedure:

Part I Activity
1. As you work through this activity, independently, be sure to write down areas that may be confusing. It is important to address any misunderstandings/misconceptions you may have before starting Part II.
2. The first problem in this activity requires you find characteristics of data (mean, median, mode, and range) when given a frequency chart. As you work through this problem, here are some things to think about:
   a. What does frequency mean?
   b. How do you calculate mean, median, mode, and range?
   c. If you are struggling with understanding the chart, is there a way to rewrite the data as a list of numbers so that you can easily calculate the mean, median, mode, and range?
3. The second problem requires the same characteristics, but now you are given a frequency line graph. Read the background information of the problem. Here are some suggestions as you work through this problem:
   a. What is the problem about (i.e. what is the context)?
   b. The graph has two axes. What do the axes represent in the context of the problem? The term "frequency" refers to the number of what in the context of the problem?
   c. What does the nature of the graph tell you about how students spend their money per month on their cell phones?
4. The third problem is similar in nature, but give a frequency bar graph.

    a. How is this graph similar/different than the graph given in problem #2?

    b. How many students participated in the goal shoot-out?

    c. Is there another way to write the data (i.e. as a list) so you can calculate the characteristics of data more easily?

Part II Activity

For this partner activity, determine which frequency graph matches to a frequency table and characteristics chart. (You may need to cut out the pieces before you begin). You should have four matches in total. You are free to create the matches in any order you like, however, you are encouraged to write down your reasoning. Notice that the characteristics data set has some missing numbers. You are required to fill in those missing pieces as well as create your matches. As you work through this activity, you and your partner will have to agree on which matches go together and will need to explain your process to each other. You will also be sharing your explanations with other students in the classroom, so be sure you are clear in your responses!

Discussion:

1. Part II – Take this opportunity to discuss with your peers how they came to their specific answers, especially if they differ from yours.

2. Part III – This activity requires you to think about the data more holistically. If you are still struggling in how to read the graphs, discuss with a partner on how you can calculate the characteristics of data (similar to problem #2 in Part I Activity). How does a data display like this inform the teacher about his/her exam?

Name:_____   Date:_____

Interpreting Data Displays (Part I)

Complete the following task individually.

| 1) Given the following frequency table, | 2) The following frequency graph shows the monthly spending of a group of students on their cell phones. |

1) Given the following frequency table,

Score	Frequency
1	1
2	1
3	3
4	1
5	4
6	5
7	6
8	5
9	3
10	1

Find each of the measures below and explain (in words) how you came to your answer.

	Explanation
Mean:	
Median:	
Mode:	
Range:	

2) The following frequency graph shows the monthly spending of a group of students on their cell phones.

Find each of the measures below and explain (in words) how you came to your answer.

	Explanation
Median:	
Mode:	
Range:	
Most students spend _____ per month	

3) The bar chart represents the outcome of a penalty shoot-out competition. Each person in the competition was allowed six shots at the goal. The graph shows, for example, that four people only scored one goal with their six shots. Find each of the measures below and explain (in words) how you came to your answer.

	Explanation
Mean:	
Median:	
Mode:	
Range:	
How many people were involved in the shoot-out?	

Interpreting Data Displays (Part II)

Card Set A – Frequency Graphs

Frequency Graph A

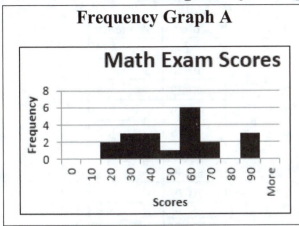

Frequency Graph B

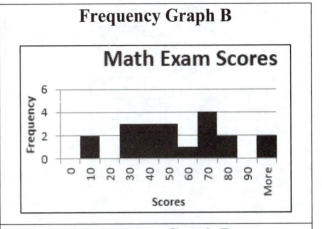

Frequency Graph C

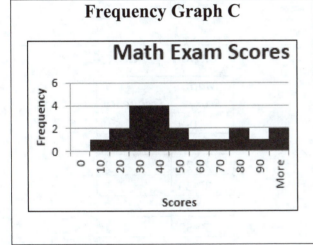

Frequency Graph D

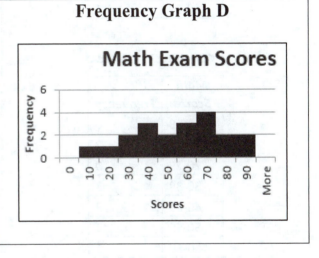

Card Set B – Frequency Tables

Frequency Table 1

Math Scores	Frequency
0-9	0
10-19	1
20-29	2
30-39	4
40-49	4
50-59	2
60-69	1
70-79	1
80-89	2
90-99	1

Frequency Table 2

Math Scores	Frequency
0-9	0
10-19	1
20-29	1
30-39	2
40-49	3
50-59	2
60-69	3
70-79	4
80-89	2
90-99	2

Frequency Table 3

Math Scores	Frequency
0-9	0
10-19	2
20-29	0
30-39	3
40-49	3
50-59	3
60-69	1
70-79	4
80-89	2
90-99	0

Frequency Table 4

Math Scores	Frequency
0-9	0
10-19	0
20-29	2
30-39	3
40-49	3
50-59	1
60-69	6
70-79	2
80-89	0
90-99	3

Card Set C – Characteristics of Data Charts

Characteristic of Data E

Mean	
Median	47
Mode	
Range	97

Characteristic of Data F

Mean	
Median	38
Mode	none
Range	

Characteristic of Data G

Mean	48.75
Median	
Mode	84
Range	

Characteristic of Data H

Mean	51.4
Median	
Mode	
Range	39

Interpreting Data Displays (Part III) – Extension Activity

Each frequency graph below represents scores of different classes on a mathematics exam. Using your knowledge of mean, median, mode, and range to give an **explanation** of how students performed on the exam in the Interpretation column.

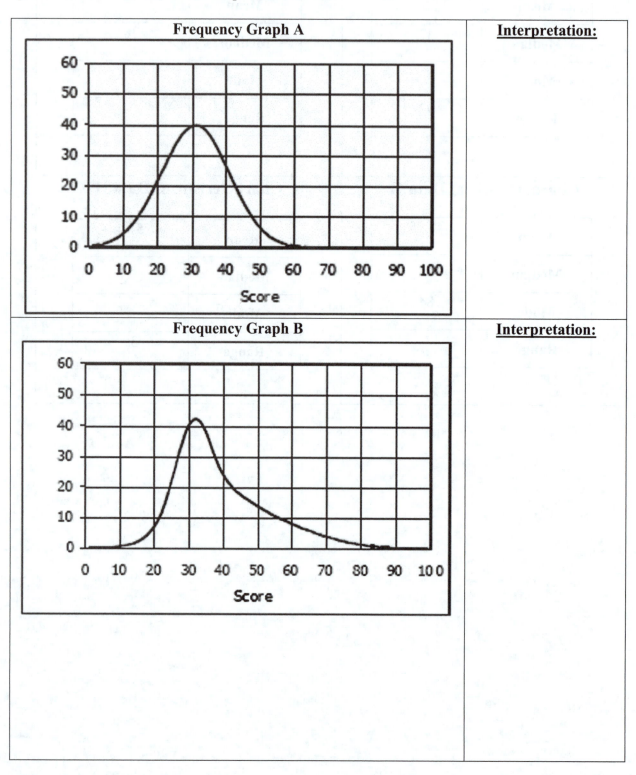

Frequency Graph A | **Interpretation:**

Frequency Graph B | **Interpretation:**

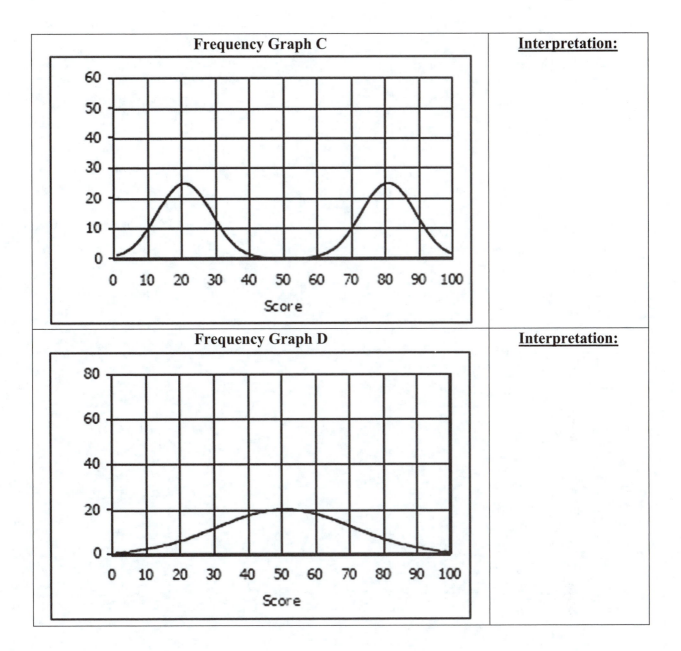

Frequency Graph C	Interpretation:

Frequency Graph D	Interpretation:

Linear Modeling and Correlation
A Case Study of Ted Williams and the War Years

Before you begin:
- Read about linear modeling in section 9-B and about correlation in section 5-E of your text.
- With a brief Internet search, become familiar with some of Ted Williams' life history.

Download Spreadsheet: (for student)
- Download or ask your instructor for the file "TedWilliams_Sheet.xls" providing you a spreadsheet framework for the case study. This spreadsheet is comprised of a completed worksheet on Ted Williams' career Home Runs and the same structure for his career Runs Batted In (RBI) but without the actual RBI historical data provided.

Procedure:
Background

    Ted Williams was the last player in the major leagues to ever bat a .400 average in a season (i.e., 40% of "at bats" resulting in hits) and he did this back in 1941. Ted Williams completely or partially missed five professional seasons during the years 1943-1945, 1952, and 1953 because of his military service as a pilot in the U.S. Air Force during World War II and the Korean War. In this case study, linear modeling (by way of statistical regression) is used to project his baseball statistics had Williams not have had any missing or partial seasons due to the wars. You will primarily address the question of whether he would have had the career record for Runs Batted In (RBI) had he played full seasons during the missing or partial five seasons. (Note that the two partial seasons during the Korean War are not included in the historical data in order to allow you the opportunity to project them as full seasons using your linear model.

Home Run Worksheet
1. Open the spreadsheet file "TedWilliams_Sheet.xls" provided by your instructor.
2. Notice the scatterplot showing the career trend of home runs hit by Ted Williams and the trend line running through the data points (sometimes referred to as a "line of best fit" or a "least-squares regression line"). Notice the linear equation in the upper right corner along with the R^2, which is referred to as the coefficient of determination.
3. Explore several of the formula cells to figure out how the spreadsheet is making the calculations. As examples, notice cells C26 through C30 contain the function '=forecast' which your spreadsheet uses to statistically project the missing data points by way of a linear regression model. Notice also that cell F30 contains the correlation coefficient based on the historical data and enables us to quantify how well the data fit to a line.
4. Pay close attention to the two different projections in cells F27 and J27. These projections represent two different techniques of accurately projecting Ted Williams' performance had he have played full seasons during the war years. Cell

F27 takes his home run average and multiplies it by the missing five years to then add to the historical total. Cell J27, however, adds the historical total to the sum of the linear model projections for the missing five years.

5. After exploring the home run worksheet, click on the RBI worksheet and notice how the historical RBI data is missing and therefore the scatterplot and analyses are incomplete.

Researching Historical Data from Major League Baseball
1. Visit the website www.mlb.com, click on the Stats menu, and then select Historical.
2. Scroll down the website and look for the Career button to select on the left-hand side. Click Career and select from the pull-down menu All Time. This allows you to research any of the all-time career records of anyone who has played or is currently playing.
3. Now select the Home Runs column (HR). For the Analysis section below, make a note of who has the record and how many. Likewise, make a note of how many career home runs Ted Williams achieved and his position number in the standings.
4. Now select the Runs Batted In column (RBI). Note the record holder's performance as well as that of Ted Williams as you did previously with home runs.
5. To finish your RBI research using this website, click on Ted Williams' name and make a note of his seasonal RBI performances.
6. Enter these RBI seasonal totals in your Historical Data Set of the RBI worksheet, remembering to leave out the partial seasons during the Korean War.

Analysis: With the help of your spreadsheet, complete the table below.

Research	Home Runs (HR)	Runs Batted In (RBI)
All-time leader (Who and how many?)		
Ted Williams (Record standing and how many?)		
Ted Williams Theoretical Total (Average method)		
Ted Williams Theoretical Total (Linear model method)		
Ted Williams Historical Correlation Coefficient		

Discussion:
1. Based on your two different scatter plots (Home Runs versus RBI), which of the two baseball statistics of Ted Williams seems to fit a line the best?
2. How do the home run and RBI correlation coefficients support your opinion?
3. Theoretically, had Ted Williams have played for the full five seasons during the war years, how many home runs would he have had and would this total have been a record?
4. What is the slope of the home run trend line in the scatter plot and how is it interpreted?

5. Theoretically, many RBI would Ted Williams have had and would this total have been a major league baseball record?
6. What is the slope of the RBI trend line in the scatter plot and how is it interpreted?
7. If you had to choose between the average method of projecting performance versus the linear modeling method, which would you use and why?
8. Suppose we define the independent variable of this case study to be Time (the years Ted Williams played major league baseball) and the dependent variable to be either Home Run or RBI performance. Therefore *interpolation* is the process of making statistical projections "within the data set". More specifically we mean within the independent *x*-values (Time). On the other hand, *extrapolation* is the process of making statistical projections "outside the data set", that is, outside the independent *x*-values (Time). Which of these two types of projections (interpolation versus extrapolation) is this case study based upon and why?
9. Which of the two, (interpolation or extrapolation), would you consider to be more risky in terms of its accuracy and why?
10. If time and opportunity permit, try to find out the role that R^2 plays in statistical studies.

Mean Versus Median

Before you begin:
- Review how to calculate mean, median, and mode.

Background: As described by your readings, the term "average" is often used in multiple contexts and does not necessarily mean taking the sum of your data and dividing by the number of items in the dataset. Sometimes, the term "average" can be used to refer to other measures of center, such as the median or the mode. In this activity, we will explore how the various meanings of "average" and determine which measure of center would better represent the data.

In this activity you will:
1. Calculate mean, median, and mode of various data.
2. Explore the measures of center (mean, median, and mode) and determine which "average" (mean or median) would be a better representation of data given.
3. Understand that data can be skewed.
4. Apply your knowledge to real-life problems and explain how a single number can represent the nature of data.

Procedure:
1. As you work through the Part I Activity, here are some guided questions to help you in your endeavor.
   a. For problem #1, determine who is correct (Sarah or Andrew) in giving the average price for a particular energy bar. These are some questions you should be asking yourself:
      i. How did Sarah calculate her answer?
      ii. How did Andrew calculate his answer?
      iii. Who do you believe is correct and why?
      iv. In this particular scenario, what does the term "average" mean?
      v. The price difference between Sarah's calculations and Andrew's calculations may not be much, but can you think of an instance where it would matter which "average" is used?
   b. For problem #2, determine how Mrs. Smith's class performed on a particular quiz.
      i. Part A: <u>Before making any calculations</u>, look at the data overall and give your best educated guess as to how you think the students did on the quiz. Which "average" do you think would best describe her students' performance?
      ii. Part B: Based on your observations in Part A, calculate the necessary "average" you think will provide Mrs. Smith with the best feedback. Which "average" did you choose and why? Now that you see the number that represents the data, do you agree with your initial assessment in Part A?

   c. For problem #3, explain why the statement made by the college is misleading. Here are some guiding questions to help you along the way:

      i. Part A:

1. Just by looking at the data, what do we know about the 5 basketball players? Did they all receive a contract?
2. The college claimed that "the average senior on this basketball team received a $2 million contract offer." Which "average" are they referring to? Do you agree/disagree with their statement? Why or why not?

     ii. Part B:

1. How else could we calculate the "average"? Why would this "average" be a better representation of the data?
2. How does the $10,000,000 affect the dataset as a whole? What kinds of numbers drastically affect the dataset?

Discussion:

1. As with the Part I Activity, determine which "average" would be a better fit for the data given. Notice that the first two scenarios are very similar to those done in the Part I activity. Given a dataset, calculate and determine whether the mean or median would be a better representation of the data. As you work through these two problems, be sure to calculate BOTH the mean and median. Be careful in how you choose which "average" to use since the question asks for a particular value.

2. For the second part of this activity, determine which "average" would be a better representation WITHOUT being given a specific data set. This will require you to think about WHO is requesting or wants the data and then determine which "average" would better suit their needs. In real-life settings, most companies like to portray themselves in a better "light," so you will have to think critically about how best to do that. Try to think of all the possibilities that can occur and if you need to, "create" a data set to help you determine which "average" to choose.

Name:_____   Date:_____

Mean Versus Median (Part I)

1) Sarah and Andrew were comparing prices of their favorite energy bar. Eight grocery stores sell the PR energy bar for the following prices:

$1.09 $1.29 $1.29 $1.35 $1.39 $1.49 $1.59 $1.79

Sarah claims the average price of the candy bar is $1.37 but Andrew disagreed and said the average price of the energy bar is actually $1.41. How did Sarah and Andrew come up with these prices? Based on their calculations, who do you think is correct and why?

2) Ms. Smith, a math teacher, recently gave a mathematics quiz in her class. The ten quiz scores were:

89   87   93   90   12   91   88   87   83   91

a) Based on the test scores above, would you say the class did well? Why or why not?

b) If you were Ms. Smith, which average would you use to describe the data: mean, median, or mode?

3) Suppose that five graduating seniors on a college basketball team receive the following first-year contract offers to play in the National Basketball Association (zero indicates that the player did not receive a contract offer):

| 0 | 0 | 0 | 0 | $10,000,000 |

The college claimed that the average senior on this basketball team received a $2 million contract offer.

a) Explain how the college came up with this number and why this statement may be misleading.

b) Would another measure of central tendency be a better representative of the data? Support your answer.

Name:_____   Date:_____

Mean Versus Median (Part II – Extension Activity)

For each of the following scenarios, determine whether the mean or median better represents the data (place a check mark in the appropriate box). For each case, explain why you chose that particular average.

Scenario	Mean	Median	Explanation
A retail store had total sales of $436, $650, $530, $500, $650, $489, and $423 last week. Which measure of data would make the store's sales last week **appear the most profitable?**			
Suppose you have opened some Nutty Bars to check the company's claim of an "average" of 8 peanuts per bar. Here is what you found after opening 10 bars: 5, 8, 8, 8, 11, 7, 8, 6, 6, and 6. Which average should the company use to **support their claim?**			

The following three scenarios below do not have a specific data set. Be sure to consider all possibilities/outcomes! "Create" a data set if you need to.

Scenario	Mean	Median	Explanation
The average number of pieces of lost luggage per flight **from an airline company's perspective**			
The average weight of potatoes in a 10-pound bag			
The average age at first marriage for men in America			

Hypothesis Testing and Confidence Intervals with M&Ms

Before you begin:
- Purchase M&Ms of the same variety so that you have 50 M&M pieces. You can use any <u>one</u> of the following varieties (milk chocolate, peanut, peanut butter, dark, almond, or kids minis)
- You will need the "Standard Scores and Percentiles for a Normal Distribution" table (from section 6C)

Technology Notes:
A graphing or scientific calculator will be needed

Background:
This project will have you conduct a hypothesis test and create a few confidence intervals. You should be familiar with at least writing the null and alternative hypothesis, stating the conclusion, and creating confidence intervals using the margin of error formula. You should also be familiar with how to use the "Standard Scores and Percentiles for a Normal Distribution" table.

Procedure:
1. Pick a variety of M&M that you are going to use for this project. Then pick a color of M&M that you want to use for the project. The choices are blue, orange, green, bright yellow, red, and brown.
2. Now after buying the M&Ms, randomly pick 50 of those M&Ms that you bought. **Make sure you pick exactly 50.**
3. Count the number of the M&Ms of the color that you specified in problem 1. What percentage of your 50 M&Ms have that color?

   Percentage $\hat{p}$ = _____

4. Compare your answer from step 3 with the percentages provided in the table below. If they are the same percentages, then you **MUST pick a different color**. Which percentage is larger, the percentage from your sample (which we will call $\hat{p}$) or the one that Mars, Inc. provided (which we will call p)?

Mars, Inc. published the percentages of colors for M&Ms CHOCOLATE CANDIES provided below:
M&Ms MILK CHOCOLATE: 24% blue, 20% orange, 16% green, 14% bright yellow, 13% red, 13% brown.
M&Ms PEANUT: 23% blue, 23% orange, 15% green, 15% bright yellow, 12% red, 12% brown.
M&Ms KIDS MINIS: 25% blue, 25% orange, 12% green, 13% bright yellow, 12% red, 13% brown.
M&Ms DARK: 17% blue, 16% orange, 16% green, 17% bright yellow, 17% red, 17% brown.
M&Ms PEANUT BUTTER: 20% blue, 20% orange, 20% green, 20% bright yellow, 10% red, 10% brown.
M&Ms ALMOND: 20% blue, 20% orange, 20% green, 20% bright yellow, 10% red, 10% brown.

5. Use the answer from step 4 to complete the following claim based on your sample: "The published percentage of [insert color and type of M&M] from Mars, Inc. should be [choose one: greater, less] than [p, the percentage that Mars, Inc. provided]."

Claim: "The published percentage of _____ M&Ms from Mars, Inc. should be _____ than _____%.

6. Use the claim from step 5 to write the null and alternative hypothesis.

7. The next step in the hypothesis test is to calculate what we call the "test statistic". The test statistic is given by the following formula. Test Statistic: $z = \dfrac{\hat{p} - p}{\sqrt{\dfrac{p(1-p)}{n}}}$

    Use this formula to calculate the test statistic z. *Note: "$\hat{p}$" is the percent from your sample, "p" is Mars, Inc.'s published percentage, "n" is the total number of M&Ms used (so $n = 50$). In the formula, both $\hat{p}$ and p need to be written as decimal numbers.*

8. Notice that the test statistic is a z-score (or a standard score). After calculating the test statistic, we can calculate the "p-value" by using the "Standard Scores and Percentiles for a Normal Distribution" table. If the alternative hypothesis says "<u>less than</u>", then the p-value is the <u>area to the left</u> of the test statistic z (so we use the percentile for that z-score). Otherwise, if the alternative hypothesis says "<u>greater than</u>", the p-value is the <u>area to the right</u> of the test statistic z (so you take 100 – percentile). Find the p-value and write it as a decimal number, not as a percentage.

Now we will look at making confidence intervals.

9. Approximate the margin of error by using the formula: $margin\, of\, error \approx \dfrac{1}{\sqrt{n}}$.

10. Use this margin of error approximation to find the 95% confidence interval.

11. Now, there is a more accurate formula for the margin of error. $E = z_{\alpha/2} \times \sqrt{\dfrac{\hat{p}(1-\hat{p})}{n}}$. For a 95% confidence interval, $z_{\alpha/2} = 1.96$. Use this new formula to calculate the margin of error "E".

12. Use this new margin of error to calculate the 95% confidence interval.

Analysis:
1. The *p*-value is defined as the probability of a sample result that is at least as extreme as the observed result. Knowing this, which of the following types of evidence do we have to reject the null hypothesis and accept the alternative hypothesis: strong, moderate, or not sufficient evidence?

2. Based on your answer, do we have sufficient evidence to support the claim mentioned in problem 4?

3. How do the two confidence intervals compare to each? Does the margin of error $\approx \dfrac{1}{\sqrt{n}}$ formula make a decent approximation for the more accurate formula?

Discussion:

1. According to the hypothesis test that you conducted, does it make sense that Mars Inc. took down their distributions from their website? Why or why not?

2. $z_{0.05}$ is the z-score for a data value (from a normal distribution) which has only 5% of the data values larger than itself (and thus 95% of the data values are less than itself). In other words, $z_{0.05}$ is the z-score that corresponds to the 95th percentile. Find $z_{0.01}$, the z-score that has only 1% of the data values larger than itself.

3. The confidence level for a confidence interval is equal to $1-\alpha$. So a 95% confidence interval will have $\alpha = 0.05$ since $0.95 = 1 - \alpha = 1 - 0.05$. Find α for a 98% confidence interval.

4. If $\alpha = 0.05$, then $z_{\alpha/2} = z_{0.05/2} = z_{0.025}$ which means that it is the z-score that corresponds to the 97.5th percentile. The closest percentile on the table is 97.72 which has the z-score of $z = 2$ (a more accurate value for $z_{0.025}$ is 1.96). Use this information and the formula

    $$E = z_{\alpha/2} \times \sqrt{\frac{\hat{p}(1-\hat{p})}{n}}$$ to find the margin of error for the 98% confidence interval.

5. What is the 98% confidence interval? How does this compare to the 95% confidence interval? Which interval is wider and why is it wider?

Weird Dice

Before you begin: If necessary, construct your Sicherman and Efron dice using blank dice and small adhesive labels, or by using the templates included at the end of this activity.

Background: You should be familiar with constructing and analyzing probability distribution tables as outlined in section 7A. You should also have some experience using the formulas $P(A \text{ and } B) = P(A) \times P(B)$ and $P(A \text{ or } B) = P(A) + P(B) - P(A \text{ and } B)$ described in section 7B

Part 1: Sicherman Dice

Sicherman Dice are a pair of six-sided dice, one (blue) with faces labeled 1, 2, 2, 3, 3, 4 and the other (green) with faces labeled 1, 3, 4, 5, 6, 8.

Procedure: In this activity, you will compare and contrast Sicherman dice with standard six-sided dice. Assume that one of your standard dice is red and the other standard die is white.
1. Construct a probability distribution table for the sum of the two dice on one roll of a standard pair of dice. This should include each possible outcome and the probability that outcome is rolled.

2. Construct a probability distribution table for the sum of the two dice on one roll of the Sicherman dice. This should include each possible outcome and the probability that outcome is rolled.

Analysis:
1. What is the probability of rolling doubles on one roll of a standard pair of dice?

2. What is the probability of rolling doubles on one roll of the Sicherman dice?

Discussion:
1. How are the standard dice and Sicherman dice similar?
2. How are the standard dice and Sicherman dice different?
3. Think of some games that are played using a pair of standard six-sided dice. How would these games be affected by using Sicherman dice?
4. Can you think of any games which could be played with the Sicherman dice instead of the standard dice without affecting the game in any way?

Part 2: Efron Dice

Efron Dice are four six-sided dice, labeled as follows.

> The red die has four faces labeled 4 and two faces labeled 0
> The green die has all six faces labeled 3
> The blue die has two faces labeled 6 and four faces labeled 2
> The yellow die has three faces labeled 5 and three faces labeled 1

Procedure: Your opponent chooses one of the four Efron dice to roll. After they select their die, you select your die from the remaining three Efron dice. You each roll your die and the higher number wins.

1. Suppose your opponent selects the red die and you pick the yellow die. Complete the following table below by placing W for a win, L for a loss, or T for a tie.

		Your Die (Yellow)					
	Die Roll	**1**	**1**	**1**	**5**	**5**	**5**
	0	W					
	0	W					
Opponent's Die (Red)	**4**	L					
	4						
	4						
	4						

2. Based on this table, what is the probability of you winning when your opponent selects red and you pick yellow? Is the yellow die a good choice for you to pick if your opponent picks the red die?

3. Suppose your opponent selects the yellow die. Which die can you select so that the probability of you rolling a higher number than your opponent is greater than 50%? With that die, what is the probability you roll a higher number than your opponent? (You may find it helpful to construct a table like the one provided in step 1.)

4. Suppose your opponent selects the blue die. Which die can you select so that the probability of you rolling a higher number than your opponent is greater than 50%? With that die, what is the probability you roll a higher number than your opponent? (You may find it helpful to construct a table like the one provided in step 1.)

5. Suppose your opponent selects the green die. Which die can you select so that the probability of you rolling a higher number than your opponent is greater than 50%? With that die, what is the probability you roll a higher number than your opponent? (You may find it helpful to construct a table like the one provided in step 1.)

Analysis:
1. If your opponent selects the red die and you pick the blue die, there are two ways you can win: either your opponent rolls 0 *or* your opponent rolls 4 and you roll 6. You can label each of these events as follows:

    event A = your opponent rolls 0
    event B = your opponent rolls 4
    event C = you roll 6

    This means that the probability you win in this situation is

    $$P(A \text{ or } (B \text{ and } C)) = P(A) + P(B \text{ and } C) - P(A \text{ and } (B \text{ and } C))$$

2. Calculate *P(A)*.

3. Calculate *P(B and C)* using the *and* probability formula for independent events.

4. Calculate *P(A and (B and C))*.

5. What is the probability you will win if your opponent rolls the red die and you choose the blue die?

Discussion:
1. The following describes a winning strategy for Efron dice.
   - If my opponent chooses the red die, I should choose the _____ die.
   - If my opponent chooses the yellow die, I should choose the _____ die.
   - If my opponent chooses the blue die, I should choose the _____ die.
   - If my opponent chooses the green die, I should choose the _____ die.
2. If you are forced to choose your die first, which die would you pick?
3. Is there a single best die to choose?

Dice Templates

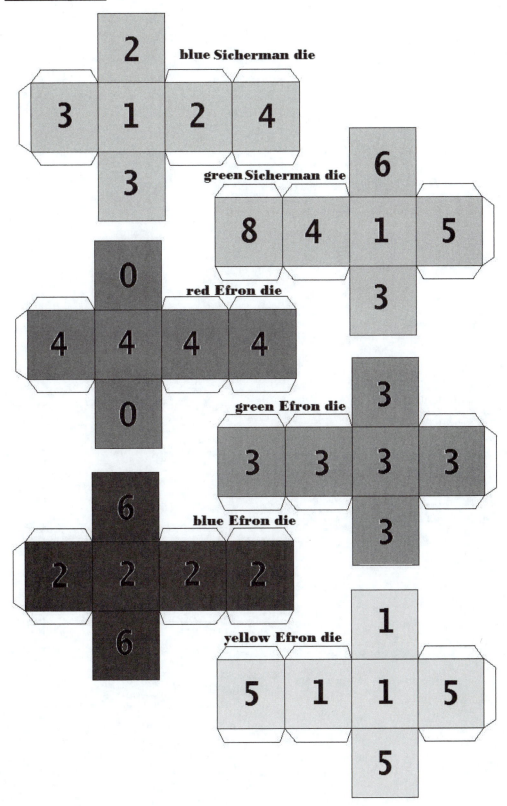

blue Sicherman die

green Sicherman die

red Efron die

green Efron die

blue Efron die

yellow Efron die

Law of Large Numbers and Potential Streaks
A Die Rolling Simulation

Before you begin: You will need a fair six-sided die (showing numbers 1 to 6) to roll and gather data for the class simulation of the Law of Large Numbers. Because part of this activity also investigates the idea of streaks, read the following sports excerpt of Shaquille O'Neal's NBA record in free-throw futility.

"There were his usual clunkers off the rim, and several other tries actually were close to the mark, but spun in and out of the hoop. When it was over, Shaquille O'Neal had broken Wilt Chamberlain's record for free-throw futility in a single game. Eleven times, O'Neal furrowed his brow, remembered to bend his knees and tried to arc the ball into [the] basket. None made it.

That gave the Los Angeles Lakers' big center, the game's most dominant player, the dubious distinction of shooting the most free throws ever in an NBA game without making a single one. Chamberlain, who also dominated the court during his day and also was notoriously bad from the line, went 0-for-10 against Detroit on Nov. 4, 1960."
(Ken Peters, *The Associated Press*, December 9, 2000)

Procedure: For the number of rolls as indicated by your instructor, record your results (numbers 1 to 6) in the table(s) below. Beneath each result (in the "Parity" row), place an X if the result was an even number (2, 4, or 6) and leave it blank for an odd result (1, 3, or 5).

Roll #	1	2	3	4	5	6	7	8	9	10	11	12	13	14	15	16	17	18	19	20
Result																				
Parity																				

Roll #	21	22	23	24	25	26	27	28	29	30	31	32	33	34	35	36	37	38	39	40
Result																				
Parity																				

Roll #	41	42	43	44	45	46	47	48	49	50	51	52	53	54	55	56	57	58	59	60
Result																				
Parity																				

Roll #	61	62	63	64	65	66	67	68	69	70	71	72	73	74	75	76	77	78	79	80
Result																				
Parity																				

Roll #	81	82	83	84	85	86	87	88	89	90	91	92	93	94	95	96	97	98	99	100
Result																				
Parity																				

Results Summary:

1. Indicate the total number times you rolled the die, the number of times you rolled a 3, and the percentage of the total number of rolls:

    total # of rolls: _____   # of rolls of a 3: _____   of rolls of a 3: _____

2. Indicate the number of rolls resulting in an even number and the percentage:

    # of even rolls:_____   % of even rolls out of total: _____

3. Now search the parity (even/odd) of your rolls and find your longest streak of either X's or non-X's.   Longest streak of even or odd numbers: _____

Analysis:

1. What type of probabilities are you finding in questions 1 and 2 of the analysis above?

    Circle one: Empirical   Subjective   Theoretical

2. How does your percentage of rolling a 3 compare to the proportion $1/6 \approx 16.7\%$?

3. Your instructor will collect and display the cumulative class results. What do you notice about the line chart of the cumulative class simulation? To what value does the percentage of rolls of the number 3 seem to be trending?

4. In your own words, try to describe in a brief sentence or two, what the Law of Large Numbers really means:

5. What was the longest streak of even or odd rolls in your class?   _____

6. How can rolling your die and recording the parity (even/odd) relate to the tossing of a fair coin or the simulating of Shaquille O'Neal's career free throw shooting?

7. By the time Shaquille O'Neal broke the record in free-throw futility (longest streak of misses), he had played for nine seasons of 82 games each, making an average of just 51.3% of his free throw shots.

    a. In light of the number of games and his overall free throw percentage, does his streak of 11 misses in a row seem surprising?

    b. How does it compare to the longest streak of evens or odds that someone had in your class?

The Monty Hall Problem

Before you begin: If necessary, review the concepts of empirical probability, theoretical probability and the Law of Large Numbers.

Background: The game show *Let's Make a Deal* featured three doors, labeled A, B and C. One door has a car behind it, while the other two doors have goats behind them.

The host of the game is named Monty Hall, and he knows which door has the car. He asks you to select a door. Let's say you select door C.

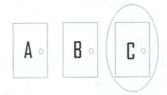

Monty then opens one of the remaining two doors, showing you a goat. He never opens the door containing the car. Let's say he opens door A.

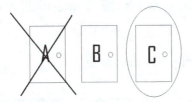

Monty then gives you the option to stick with your original door (in this example, door C) or switch to the other door that he didn't open (in this example, door B). The **Monty Hall Problem** is determining whether you should stick with your original door or switch to the other door. Do you have a better chance of winning if you switch? Do you have a better chance of winning if you stick with your original choice? Does it even matter?

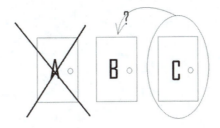

Procedure: In this activity, you will simulate the Monty Hall problem with cups and candy. You will determine the probability of winning when you stick with your original choice, and the probability of winning when you switch.

1. Why can Monty always show you a goat? Discuss this question with your partner.

2. Take turns playing "Monty" and "contestant":
   a. The "contestant" turns around while "Monty" places the candy under one of the three cups.
   b. The "contestant" turns around and selects one of the cups.
   c. "Monty" lifts one of the empty cups and asks the "contestant" if they want to stick with their original cup or switch to the other cup.
   d. The "contestant" makes their choice.
   e. "Monty" reveals the candy.
   f. The result of the simulation is tallied in the attached table of results.
   g. Repeat this game many times!

3. Some important points to remember:
   - As a pair, you should stick with your original cup about half the time and switch to the other cup about half the time.

   - Keep an accurate record of your simulations! Be sure you understand what each quadrant of the table of results represents.

   - "Monty" must not give anything away when the "contestant" makes their original choice.

   - "Monty" has to remember which cup has the candy!

   - The "contestant" should not touch any of the cups at any point in the game.

   - If anything spoils an individual round of the game, ***do not*** record that round in the table.

4. When prompted by your instructor, total up the tallies in each portion of your table of results and record these in the class's community table.

Analysis:

1. Looking at the large community table of results which summarizes the entire class's data, what is the empirical probability that a person will win when they stick with their original choice?

2. Looking at the large community table of results which summarizes the entire class's data, what is the empirical probability that a person will win when they switch choices?

3. Do these answers surprise you? Discuss with your partner and/or the entire class.

4. What is the theoretical probability that a person will win when they stick with their original choice? Devise a convincing argument that works independent of the large community table of results which summarizes the entire class's data.

5. What is the theoretical probability that a person will win when they switch choices? Devise a convincing argument that works independent of the large community table of results which summarizes the entire class's data.

6. Discuss how your answers to these analysis questions relate to the Law of Large Numbers.

Discussion:

1. Suppose you are playing a version of the Monty Hall game with 100 doors, but otherwise it is the same as described in class. One door contains a car, while the other 99 doors contain goats. You select one door, and Monty opens 98 other doors which all contain goats, leaving only two doors closed. Remember, Monty knows where the prize is and does not reveal it. You may stick with your original door, or switch to the other door. What should you do? Why?

2. Consider the "Monty from Hell" version of the Monty Hall problem. In this version, the game is played with three doors. You choose a door and Monty opens another door which does not contain the prize. However, Monty only offers you the choice to switch if he knows you picked the door *with* the prize. What should you do? Why?

3. Consider the "angelic Monty" version of the Monty Hall problem. In this version, the game is played with three doors. You choose a door and Monty opens another door which does not contain the prize. However, Monty only offers you the choice to switch if he knows you picked a door *without* the prize. What should you do? Why?

4. Consider the "ignorant Monty" version of the Monty Hall problem. In this version, the game is played with three doors, and Monty *does not know* which door contains the car. You choose a door, and Monty opens another door which just so happens to not contain the prize. You may stick with your original door, or switch to the other door. What should you do? Why?

5. Suppose you are offered a choice of one of three boxes. One box contains two gold coins, one box contains two silver coins, and one box contains one gold coin and one silver coin. You randomly select one box and remove one coin from it. The coin is gold. What is the probability the other coin in the box is also gold? (This problem is known as Bertrand's box problem.)

Table of Results

	Win	Lose
Stick with your original cup	Total	Total
Switch to the other cup	Total	Total

Understanding the Birthday Problem
A Numerical, Graphical, and Practical Investigation

Before you begin: Review the section titled *Probability and Coincidence* in Unit 7E of your textbook and especially focus on Example 8: Birthday Coincidence to get an overview of the classic birthday problem. Try to anticipate the shape of the probability curves by sketching your best guess in the rectangles below. Let the *y*-axis represent the approximate probabilities ranging 0 to 1 (that is, impossible to certain). If you are working with a partner, discuss the possible shapes first before committing to paper.

Probability of Any Birthday in Common

Probability of a Particular Birthday in Common

5 10 15 20 25 30 35 40 45 50
Number of People in a Room

100 200 300 400 500 600 700 800 900
Number of People in a Room

Procedure: Now, with a blank spreadsheet, carefully follow the steps below:

A Particular Birthday in Common

1. Type in cell A1, "Particular Birthday in Common" to act as a header for this first scenario.
2. Now in cell A3 type "Persons" and in B3 type "Probabilities."
3. In cell A4 type 50 and A5 type 100.
4. With that pattern completed (increment of 50 each time), highlight cells A4 and A5 and then fill down the pattern to 1500 persons (point the arrow in the lower right corner and drag downward on the black cross hair).
5. To calculate the corresponding probabilities, refer to Example 8 part (a) and type the following formula in cell B4: $=1-(364/365)\wedge A4$. This uses the at least once rule and cell A4 contains the number of persons in the room besides yourself.
6. Finish the probabilities by filling the formula in cell B4 down to the cell with 1500 persons.

Any Birthday in Common

1. Type in cell D1, "Any Birthday in Common" to act as your header for the second scenario.
2. Now in cell D3 type 'Persons' and in E3 type "Probabilities."

3. In cell D4 type 2 and D5 type 3.
4. With that pattern completed (increment of 1 person each time), highlight cells D4 and D5 and then fill down the pattern to 50 persons (use the black crosshair in the lower right corner) click and drag downward as before.
5. To calculate the corresponding probabilities, refer to Example 8 part (b) and type the following formula in cell E4: $=1-\text{PERMUT}(365, D4)/365\,\hat{}\,D4$. As mentioned in the text, this formula still uses the *at least once* question but cell D4 contains the number of persons in the room <u>including</u> yourself. Notice how the function PERMUT accomplishes what many calculators cannot accomplish due to the large numbers. (Only calculators that have a capacity of 10^999 can handle these calculations. Examples include the TI-89 and TI-86.)

<u>Graphs of the Probabilities</u>
1. Review the Using Technology boxes of 5C to learn how to construct a line chart or scatterplot effectively.
2. Highlight the Particular Birthday probabilities from column B and create a line chart that displays the probabilities as they vary with the number of people. Now repeat the process for the probabilities in column E for the Any Birthday in Common scenario.
3. Print your line charts with adequate titles and labels if instructed by your teacher.

Results Summary/Analysis: Now with the two types of probabilities calculated on your spreadsheet, fill in the following tables and complete the questions below. Round your probabilities in the tables below to two decimal places.

A Particular Birthday in Common:

Persons	100	200	300	400	500	600	700	800	900
Probabilities									

Any Birthday in Common:

Persons	5	10	15	20	25	30	35	40	45
Probabilities									

1. How many students are in your class counting yourself and the teacher? _____
2. What is the probability that someone in the class has your very birthday? _____
3. What is the probability that there is a birthday match (that is, some birthday in common) in the room? _____
4. How confident are you that there will actually be a match for question #3 with your particular class of students?

5. ** With your teacher's help, the class will find out whether there is a birthday match in the room. After this practical survey, describe the outcome and your thoughts or insights:

6. According to your data from your spreadsheet, how many people are needed in a group in order to have approximately 0.50 probability that someone has a birthday in common with someone else in the group? _____

7. Now answer the same question from #6 but for the "Particular Birthday" scenario. That is, between _____ and _____ people would be needed in the room to have a 0.50 probability that two people share a particular birthday. What is the most accurate answer you can find? (Modify your spreadsheet slightly to find it.) _____

Discussion:

1. How do your actual line charts compare to your first impressions you sketched out in the Before You Begin section above? Were you close in terms of shape or significantly off from your first guesses?
2. Disregarding Leap Day, how many people would be needed in the room to be 100% confident that at least two people share the same birthday? (This is known as the Pigeonhole Principle.) Justify your answer.
3. Now, looking at your results on the spreadsheet for the Particular Birthday scenario, discuss the theoretical number of people needed in a room to be absolutely certain someone has your birthday. Why is this different from your answer in question #2?
4. Describe how the two line charts (Particular Birthday vs. Any Birthday) differ in terms of the shape and the number of persons on the x-axis.
5. At approximately what probability does the curve from the Any Birthday line chart seem to bend? Or in other words, where does it change from increasing "faster and faster" to increasing "slower and slower"?
6. The story is told of a guest on the Johnny Carson's Tonight Show, back in the 1970s, who tried to impress everyone about the classic birthday problem. He told Cason that even though the studio audience consisted of only about 120 people (less than 1/3 of the number of days in the calendar year) there was an extremely high probability that someone would have Johnny Carson's birthday in the studio and the guest was obviously embarrassed. What would the probability have been for success given that there were about 120 people in the audience? Discuss how the guest was confused about the classic birthday problem and how it should have been stated on the show correctly.

U.S. Workforce Population
Where's the Growth?

Before you begin:

- This activity utilizes StatCrunch which is available to you through your MyLab Math class access.

Procedure: In recent years there has been much discussion in the political arenas regarding U.S. population growth and unemployment which impacts a wide array of arenas such the housing market. Media contains both accurate as well as biased information. In this activity a pre-assessment will test your current knowledge through estimates regarding population growth of the U.S. workforce. StatCrunch will then be utilized to explore actual data from 1948 to 2015 as reported by the US Bureau of Labor Statistics.

Pre-assessment

Individually or in groups as directed by your instructor, answer the following questions to the best of your ability without the assistance of any resources such as your text and electronic search engines:

1. Estimate the size of today's US adult (16 and older) population which is either working or looking for work.

2. Estimate the number of male and females currently not working (retired, in college, not looking for work)

3. If working in groups, share your estimate with your group and compare estimates. Describe differences between your group members and possible reasons given for the estimates.

4. For the three populations listed below, circle the term that best represents the population change from the mid 1900's to current day volume of workers.

US workforce	decreasing	constant	increasing
Male U.S. workforce	decreasing	constant	increasing
Female U.S. workforce	decreasing	constant	increasing

Analysis: Log in to your MyLab Math class where you will find access to StatCrunch in the main vertical toolbar of your class. If you do not find StatCrunch, please ask your instructor for assistance.

Follow the below steps to access the data utilized in this activity:
1. Sign-in to MyLab Math and enter your class
2. Select StatCrunch from the vertical toolbar
3. Click on "Explore" in the horizontal toolbar at the top
4. Click on "Data" under "Explore the StatCrunch site"
5. In the left vertical toolbar search for 'Population'
6. Select "US Workforce Participation"

The Bureau of Labor Statistics defines the Annual Average Workforce Participation as "the percentage of the population [16 years and older] that is either employed or unemployed (that is, either working or actively seeking work).

Review the data listed in the spreadsheet to answer the following questions:

1. Using the Year and Annual Average columns, what trend(s) do you notice?

2. Using the Year and Male Workforce Participation Rate columns, what trend(s) do you notice?

3. Using the Year and Female Workforce Participation Rate columns, what trend(s) do you notice?

The inactivity rates displayed in the spreadsheet are defined as "the proportion of the population aged 25 – 54 that is not in the labor force. Common reasons for leaving labor force: college, retirement, stay at home, can't find work and no longer try."

4. What differences do you notice in the trends of the Male and Female Workforce Participation rates?

5. What differences do you notice in the trends of the Male and Female Inactivity Participation rates?

6. List three or more items that may have impacted the changes in the male and female inactivity rates.

7. If working in groups, share your answers to question 6 with your group members. What additional items did your group suggest?

Graphical Analysis:
Follow the below steps to create a scatter graph in StatCrunch using the data from this activity:
- Click on 'Graph'
- Select 'Scatter Plot'
- *x*-variable: Select 'Year'
- *y*-variable: Select variable as indicated in below questions
- Click on 'Compute!'

1. In StatCrunch, create three different graphs using the following for the *y*-variable.
   a. Annual Average
   b. Male Workforce Participation Rate
   c. Female Workforce Participation Rate

2. Sketch the graphs created in above question #1:

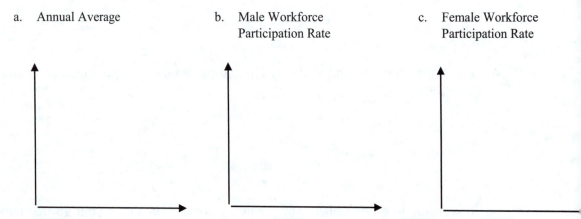

    a. Annual Average   b. Male Workforce   c. Female Workforce
                                      Participation Rate   Participation Rate

3. Describe the trends noted in the above graphs.

4. What reasons do you propose impact the changes shown in the above graphs?

5. How do the trends shown in the above graphs compare to your original thoughts shared in question #4 of the pre-assessment?

6. In StatCrunch, create three different graphs using the following for the *y*-variable.
    a. Annual Average
    b. Male Inactivity Rate
    c. Female Inactivity Rate

7. Sketch the graphs created in above question #1:

   a. Annual Average   b. Male Inactivity   c. Female Inactivity
                                Rate   Rate

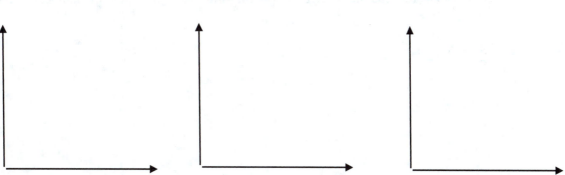

8. Describe the trends noted in the above graphs.

9. What reasons do you propose impact the changes shown in the above graphs?

10. If working in groups, share your answers to question 9 with your group members. What additional items did your group suggest?

Discussion:

1. How does the information shown in the data displayed in in the StatCrunch file align with or conflict with information shared currently in media sources?
2. Using a search engine, research and discuss items currently impacting unemployment.
3. As a class or in small groups as defined by your instructor, discuss items impacting the changing volume of the U.S. workforce.
4. Discuss reasons that may account for the opposite change rate in male and female workforce participation rates.
5. Discuss reasons that may be impacting the inactivity rates for males and females.

Are Earthquakes Becoming More Common?

Introduction:

In early 2010, two major earthquakes struck barely over a month a part, one that killed more than 200,000 people in Haiti, and another that devastated numerous communities in Chile. The closeness of the two events in time led to numerous claims that there's been an increase in the frequency of earthquakes in recent years. In this activity, you will use data from the U.S. Geological Survey (USGS) to investigate this claim.

1. The last page of this activity gives data on the number of earthquakes in various magnitude categories for the years 1970 to 2009 (data from the USGS web site). Referring to your textbook as needed, briefly describe (in a couple of sentences) what each of the magnitude categories means.

2. Look first at the number of earthquakes for each category over the period 1970 to 2009. Within your group, discuss any notable trends or exceptions that you see. For example: Does there seem to be any increase or decrease with time? Are there any years that jump out as having an unusually large or small number of earthquakes? Write down any trends or exceptions that you think may be real. Spreadsheet option: If you are using the spreadsheet with the same data, make graphs showing how the number of earthquakes in each category has changed with time.

3. Now look at the data for the estimated deaths. Do you see any trends with time? Are there any years that had surprisingly large numbers of deaths? Write down any trends or surprises that you see.

4. Overall, do you think the data support a claim that more earthquakes are occurring now than in decades past? Why or why not?

5. Should we be surprised that two major earthquakes struck only a little over a month apart in 2010, or are such occurrences to be expected once in a while?

Further Analysis/Discussion

1. Find similar data for earthquakes that have occurred since 2009, and add it to the data tables. Does 2010 seem unusual overall?

2. The tables show only earthquakes with magnitudes of 5.0 or greater. If you look up numbers for smaller earthquakes, you'll find that the number recorded has increased dramatically with time; however, scientists attribute this increase to the growth in the number and sensitivity of earthquake monitoring stations, rather than to an actual increase in the number of earthquakes. Why would the improvement in earthquake monitoring technology tend to cause an increase in the number of small earthquakes recorded, but not an increase in the number of larger earthquakes?

3. While most scientists say that no data support any claim of an increase in the number of earthquakes, most also agree that earthquakes today are causing more damage and more deaths than in the past. Why would that be the case?

4. Do some research on what areas of the world have populations that are most vulnerable to earthquakes. What can be done to help prevent massive death and destruction if and when earthquakes occur in those regions?

5. Do some research to learn about the relationship between earthquakes and tsunamis? Why are some earthquakes more likely to generate tsunamis than others?

6. What is the current status of tsunami warning systems around the world? Do you think the current systems are adequate

Earthquake Data (from USGS [www.usgs.gov])

For each magnitude range (left column), the subsequent columns show the number of earthquakes for each year. The estimated deaths are highly uncertain for larger numbers and are given to the nearest 100.

Year	1970	1970	1972	1973	1974	1975	1976	1977	1978	1979
5.0-5.9	1195	1331	1316	1331	1312	1447	1649	1686	1526	1366
6.0-6.9	110	112	110	95	99	107	114	89	93	100
7.0-7.9	20	19	15	13	14	14	15	11	16	13
>8.0	0	1	0	0	0	1	2	2	0	0
Est. deaths	68200	1300	11200	700	5400	12300	697400	2800	15200	1500

Year	1980	1980	1982	1983	1984	1985	1986	1987	1988	1989
5.0-5.9	1299	1168	1425	1673	1579	1674	1665	1437	1485	1444
6.0-6.9	105	90	85	126	91	110	89	112	93	79
7.0-7.9	13	13	10	14	8	13	5	11	8	6
>8.0	1	0	0	0	0	1	1	0	0	1
Est. deaths	8600	5200	3300	2400	200	9800	1100	1100	26600	600

Year	1990	1990	1992	1993	1994	1995	1996	1997	1998	1999
5.0-5.9	1617	1457	1498	1426	1542	1318	1222	1113	979	1104
6.0-6.9	109	96	166	137	146	183	149	120	117	116
7.0-7.9	18	16	13	12	11	18	14	16	11	18
>8.0	0	0	0	0	2	2	1	0	1	0
Est. deaths	52100	3200	3900	10100	1600	800	600	3100	9400	22700

Year	2000	2001	2002	2003	2004	2005	2006	2007	2008	200911
5.0-5.9	1344	1224	1201	1203	1515	1693	1712	2074	1768	1700
6.0-6.9	146	121	127	140	141	140	142	178	158	142
7.0-7.9	14	15	13	14	14	10	9	14	12	16
>8.0	1	1	0	1	2	1	2	4	0	1
Est. deaths	200	21400	1700	33900	228800	82400	6600	700	88000	1800

The Golden Mean

Before you begin:

Read Unit 11C of the textbook for background on the long history of the golden mean and the many ways in which it appears in the world around us.

Procedure

1. Sit back, clear your thoughts, find a receptive state of mind, and consider the following rectangles.

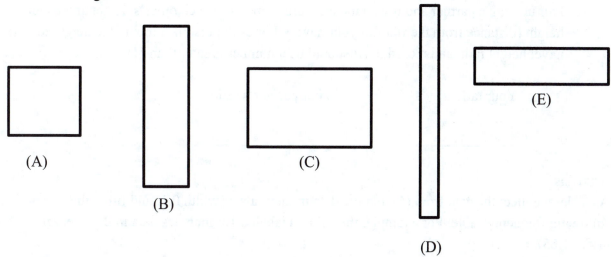

Using whatever criteria you choose, which rectangle (A–E) do you find most visually or aesthetically appealing?

Circle one choice: A   B   C   D   E

2. As a class (for example, on a whiteboard), make a frequency table and a bar chart that displays the preferences of your class in (1). What was the most common choice?

3. Measure the side lengths of each of the rectangles, and calculate the ratio of the length to the width for each one. Which rectangle has a length-to-width ratio closest to the *golden mean*, which has the value $\frac{1+\sqrt{5}}{2} \approx 1.62$? Does your class preference support or not support the claim that rectangles with proportions close to the golden mean are the most aesthetically pleasing?

Rectangle	A	B	C	D	E
Frequency					

4. Which has proportions closer to the golden mean, a standard 8.5″ × 11″ × sheet of paper or a legal-size 8.5″ × 14 ″ sheet of paper? Show your work.

5. Identify three *specific* everyday objects that have proportions close to the golden ratio.

6. Pair up with a partner and use a tape measure to measure each other's height and navel height (distance from the floor to your navel). For each person, compute the height to navel height ratio and record it. (It should be a number greater than 1!)

   Your ratio   Your partner's ratio

   _____   _____

Analysis

As a class, collect the data from (6), bin the data in intervals of width 0.1, and fill in the following frequency table. For example, the column labeled 1.6 includes data in the interval (1.55, 1.65).

Ratio	<1.15	1.2	1.3	1.4	1.5	1.6	1.7	1.8	1.9	2.0	>2.05
Frequency											

Discussion

An old hypothesis claims that the mean body height to navel height ratio of humans equals the golden mean. Do the data from your class appear to support this hypothesis? Discuss reasons that your class data might depart from this hypothesis.

Worksheets for
Using & Understanding Mathematics
with Integrated Review

Chapter 1 Thinking Critically

Learning Objectives
Objective 1 – Compare real numbers Objective 2 – Evaluate an algebraic expression Objective 3 – Classify real numbers

Objective 1 – Compare real numbers

In mathematics, we study statements - sentences that are either true or false but not both. For example, 8 is an even integer (true statement) and 10 is an odd integer (false statement).

The purpose of this section is to use integer relationships and inequalities to advance our logical reasoning skills. Our understanding of these integer relationships combined with inequality interpretation will help us recognize logical arguments that are either well supported or poorly supported by facts or assumptions (integer facts). To help analyze and interpret inequalities, we need to know the meaning of the following inequality symbols:

> **> indicates greater than**
> **< indicates less than**
> **≥ indicates greater than or equal to**
> **≤ indicates less than or equal to**

Guided Examples   Practice

1) Is the statement $5 \geq 8$ true or false (a fallacy)? The inequality symbol communicates that 5 is greater than or equal to 8. This statement is **false (a fallacy)** since 5 has a unit value less than 8. 2) Is the statement $-3 < 0$ true or false (a fallacy)? The inequality symbol communicates that -3 is less than 0. This statement is **true** since -3 has a unit value less than 0. 3) Is the statement $15 \geq 15$ true or false (a fallacy)? The inequality symbol communicates that 15 is greater than or equal to 15. This is **true** since 15 has a unit value that is equal to 15.	1) Is the statement $-3 \leq 5$ true or false (a fallacy)? 2) Is the statement $6 > 3$ true or false (a fallacy)? 3) Is the statement $12 > 12$ true or false (a fallacy)? 4) All the Presidents of the United States have been male so men are better qualified than women to be President (true or false – a fallacy).

Name:_____Date:_____

Instructor:_____Section:_____

Objective 2 – Evaluate an algebraic expression

In this section, we will evaluate an algebraic expression to aid in our understanding and mastery of propositions and logical connectors.

Proposition review

A proposition is a declarative statement that is either true or false (not both). We will explore algebraic concepts to improve our understanding of these logic principles.

For example, $8 + 12 = 20$ is a proposition (a statement) that is true.

Another logic concept is the negation. It transforms a proposition into its opposite truth value (true to false and false to true). For our algebraic example, we know $8 + 12 = 20$. The negation of this example is $8 + 12 \neq 20$.

Guided Examples	Practice
1) Is the statement $12 - 4 = 8$ true or false? This statement is true. 2) Write the negation of $8 + 2 = 10$. A negation makes the opposite claim. The negation of is $8 + 2 \neq 10$.	1) Is the statement $9 + 3 = 11$ true or false? 2) Write the negation of $6 + 4 = 10$.

Logic connector review

Logical connectors are used to join or connect two ideas. These two ideas or statements can be joined together with the connector AND. This creates a compound statement that is **true only when both parts** of the compound statement are true.

The other type of logical connector we will review is OR. This connector also joins two ideas or statements together into a compound statement. This compound statement is **true when either one or both parts** of the compound statement are true.

Guided Examples	Practice
Determine if the compound statement is true or false. 3) $3 + 9 = 12$ AND $3 \cdot 5 = 12$. Is this entire statement true or false? Since we use the connector AND, both parts of the compound statement need to be true for the entire statement to be true. $3 + 9 = 12$ is true and $3 \cdot 5 = 12$ is false, so this entire statement is false.	Determine if the compound statement is true or false. 3) $5 - (-3) = 8$ AND $2 \cdot 8 = 16$ 4) $-6 \cdot -2 = 12$ AND $(4 \cdot 7) - 10 = 14$ 5) Green beans are vegetables AND oranges are fruit.

4) $4+9=15$ OR $-5 \cdot -3 = 15$. Is this entire statement true or false?

Even though $4+9=15$ is a false statement, since $-5 \cdot -3 = 15$ is a true statement, so the entire compound statement is true.

6) $3 \cdot 2 = 6$ OR $10+10=0$

7) $5 \cdot 5 = -25$ OR $7-1=8$

8) Indiana is a state or California is a country.

Name:_____Date:_____

Instructor:_____Section:_____

Objective 3 – Classify real numbers

There are a number of different ways to classify numbers. This section will introduce the most common classifications of real numbers so you can explore, apply and master logic concepts. Specifically, we will investigate real number classification as it applies to set relationships and set theory, as well as, Venn diagram design and construction.

Natural Numbers: The natural numbers start with one and compose the numbers $1, 2, 3, 4...$ and so on. Zero is not considered a "natural number."

Whole Numbers: The whole numbers start with zero and compose the numbers $0, 1, 2, 3, 4,...$ and so on (the natural numbers including zero).

Integers: The integers are the whole numbers and their opposites (the positive whole numbers, the negative whole numbers, including zero). Integers contain $...-3, -2, -1, 0, 1, 2, 3...$

Rational Numbers: The rational numbers include all the integers, plus fractions that consist of two irrational numbers (that can be written as a quotient of integers with the denominator $\neq 0$), or terminating decimals and repeating decimals. Rational numbers contain $...-3, -2.5, -2, -1, 0, \frac{1}{3}, \frac{1}{2}, 1, 2, 3...$

Irrational Numbers: An irrational number is a number with a decimal that neither terminates nor repeats. For example, the $\sqrt{2}$ is a decimal that does not repeat itself, but that continues infinitely. The value π (pi) is classified as an irrational number as well.

Real Numbers: All the rational numbers and all the irrational numbers together form the real numbers.

This narrative description of real number classification is visually demonstrated using the Venn diagram below.

Guided Examples	Practice
1) Identify the set(s) that describe -5. Since -5 is an integer, then it is a member of the set(s): integer, rational, and real.	Identify the specified set. 1) Identify the set(s) that describe $\frac{3}{4}$. 2) Identify the set(s) that describe 7. 3) I can buy an irrational amount of apples (True or False). 4) I can buy an amount of apples that is a natural number (True or False).

Chapter 2 Approaches to Problem Solving

Learning Objectives

Objective 1 – Multiply and divide fractions
Objective 2 – Write a number in scientific or standard notation

---

Objective 1 – Multiply and divide fractions

Dimensional analysis is a mathematical system using conversion factors to move from one unit of measurement to a different unit of measurement. Typically, this involves the multiplication of two or more fractions using a specific dimensional analysis format. Our emphasis is on the precise application of multiplying and dividing fractions to become skilled at dimensional analysis calculations and develop an accurate interpretation of the data.

Multiply fractions review

There are four steps to multiplying two fractions:

- multiply the two numerators (top numbers) together
- multiply the two denominators (bottom numbers) together.
- The resulting new numerator is the product of the numerator in the first fraction and the numerator in the second fraction. The resulting new denominator is the product of the denominator in the first fraction and denominator in the second fraction.
- If need be, simplify the resulting fraction.

Guided Examples   Practice

Multiply the fractions:	Multiply the fractions
1) $\dfrac{6}{1} \cdot \dfrac{3}{5} =$	1) $\dfrac{2}{3} \cdot \dfrac{5}{7} =$
$\dfrac{6 \cdot 3}{1 \cdot 5} = \dfrac{18}{5}$	
Multiply the numerators (top numbers), then multiply the denominators (bottom numbers). The resulting fraction is your answer.	
2) $\dfrac{3}{7} \cdot \dfrac{2}{5} =$	2) $\dfrac{3}{1} \cdot \dfrac{2}{5} =$
$\dfrac{3 \cdot 2}{7 \cdot 5} = \dfrac{6}{35}$	
Multiply the numerators (top numbers), then multiply the denominators (bottom numbers). The resulting fraction is your answer.	

Name:_____ Date:_____

Instructor:_____ Section:_____

Divide fractions review

There are three steps to dividing two fractions:

- Modify the second fraction first by turning it upside-down (the reciprocal)
- Change the operation from division to multiplication then multiply the first fraction by the reciprocal (multiply across – numerator to numerator and denominator to denominator)
- If need be, simplify the resulting fraction.

Guided Examples   Practice

Divide the fractions:	Divide the fractions:
3) $\dfrac{4}{5} \div \dfrac{7}{3} = \dfrac{4}{5} \cdot \dfrac{3}{7} = \dfrac{4 \cdot 3}{5 \cdot 7} = \dfrac{12}{35}$ You must modify the second fraction. $\dfrac{7}{3}$ becomes $\dfrac{3}{7}$. Next, change the operation from division to multiplication. Multiply the first fraction and the modified second fraction Multiply across the numerators (top numbers) and the denominators (bottom numbers). 4) $\dfrac{2}{11} \div \dfrac{5}{3} = \dfrac{2}{11} \cdot \dfrac{3}{5} = \dfrac{2 \cdot 3}{11 \cdot 5} = \dfrac{6}{55}$ You must modify the second fraction. $\dfrac{5}{3}$ becomes $\dfrac{3}{5}$. Next, change the operation from division to multiplication. Multiply the first fraction and the modified second fraction Multiply across the numerators (top numbers) and the denominators (bottom numbers).	3) $\dfrac{2}{7} \div \dfrac{1}{5} =$ 4) $\dfrac{4}{1} \div \dfrac{3}{5} =$

Objective 2 – Write a number in scientific or standard notation

Scientific Notation was developed in order to easily represent numbers that are either very large or very small. To simplify matters when describing, writing or calculating with very large or very small numbers, we often use scientific notation to represent these numerical extremes.

This learning objective will focus on converting between scientific and standard notation in order to more accurately solve dimensional analysis challenges.

Converting from standard notation to scientific notation review

A number is in scientific notation when it is broken up as the product of two parts. The general form of scientific notation is $a \times 10^n$ where a is greater than or equal to 1 and less than 10, and n is an integer. For example, the value 4.7×10^3 is written in scientific notation where 4.7 is the coefficient and 10^3 is the power of ten.

There are three elements to consider when converting a number from standard notation to scientific notation:
- Place the decimal point such that there is **one nonzero digit to the left of the decimal point**.
- Count the number of decimal places the decimal has "moved" from the original number. This will determine the absolute value of the exponent of the 10.
- If the original number is less than 1, the exponent is negative; if the original number is greater than 1, the exponent is positive.

Guided Examples   Practice

Convert from standard notation to scientific notation	Convert from standard notation to scientific notation
1) $2,350,000 = 2.35 \times 10^6$ Since the decimal point is placed after the zeros, it is move to between the 2 and 3, you get a coefficient of 2.35. The decimal point has moved 6 decimal places so the power of 10 is 10^6. The original number is greater than 1, so the exponent is positive. 2) $2116 = 2.116 \times 10^3$ 3) $0.00082 = 8.2 \times 10^{-4}$ Since the decimal point is before the zeros, it is moved to between the 8 and 2, you get a coefficient of 8.2. The decimal point has moved 4 decimal places so the power of 10 is 10^4. The original number is less than 1, so the exponent is negative. 4) $0.0019 = 1.9 \times 10^{-3}$	1) $328,000 =$ 2) $14,629 =$ 3) $0.0034 =$ 4) $0.0000058 =$

Converting from scientific notation to standard notation review

There are two elements to consider when converting a number from scientific notation to standard notation:
- Move the decimal point to the right for positive exponents of 10. The exponent tells you how many places to move. The positive exponent indicates the resulting number (answer) is a large number (greater than 1).
- Move the decimal point to the left for negative exponents of 10. The exponent tells you how many places to move. The negative exponent indicates the resulting number (answer) is a small number (less than 1).

Guided Examples   Practice

Convert from scientific notation to standard form	Convert from scientific notation to standard form
5) $7.21 \times 10^4 = 72,100$ Because the exponent is positive, move the decimal point to the right 4 (the exponent value) places. The positive exponent indicates the answer is a large number (greater than 1). 6) $4.358 \times 10^{-3} = 0.004358$ Because the exponent is negative, move the decimal point to the left 3 (the exponent value) places. The negative exponent indicates the answer is a small number (less than 1).	5. $6.82 \cdot 10^3 =$ 6) $7.1 \cdot 10^5 =$ 7) $9.11 \cdot 10^{-4} =$ 8) $2.185 \cdot 10^{-2} =$

Chapter 3 Numbers in the Real World

Learning Objectives
Objective 1 – Simplify fractions
Objective 2 – Multiply and divide by decimal values to introduce percentages
Objective 3 – Percent conversions
Objective 4 – Multiply or divide using scientific notation
Objective 5 – Decimal rounding

Objective 1 – Simplify fractions

In this chapter, your challenge will be to determine, explore and interpret ratios. A ratio shows the relative size of two (or more) values. Ratios can be communicated in different forms using the example as follows:

Your survey your class and 7 of your colleagues wear glasses out of the 20 people in your class. Determine and write the ratio of people who wear glasses to the total class. Your answer can be written in several forms as follows:

- 7 : 20 (7 people wear glasses in your class of 20)

- $\dfrac{7}{20}$ (7 people wear glasses in your class of 20)

- 0.35 or 35%$(7 \div 20)$ of the people in your class wear glasses

At times, a ratio will need to be reduced to its simplest form using division. For example, you survey the 20 people in your class. It turns out that 8 of your colleagues are over the age of 25. Determine and write the ratio of people over the age of 25 to the total people in your class. Your answer can be written as

8 : 20 or $\dfrac{8}{20}$ or 40%$(8 \div 20)$. Because the ratio written in fraction form is not written in simplest form, we need to

simply the ratio. This is a two-step process that requires you to do the following:

- Find the Greatest Common Factor (GCF) of the numerator and denominator
- Divide the numerator and the denominator by the GCF

In this case, 4 is the largest number (GCF) common to both 8 and 20. We can now determine the ratio in simplest form $\dfrac{8}{20} = \dfrac{8 \div 4}{20 \div 4} = \dfrac{2}{5}$.

Guided Examples	Practice

Reduce the ratio	Reduce the ratio
1) $\dfrac{24}{32}$ $\dfrac{24}{32} = \dfrac{24 \div 8}{32 \div 8} = \dfrac{3}{4}$ Identify the Greatest Common Factor (GCF) common to both the numerator and denominator. The GCF is 8. Next, divide both the numerator and denominator by the GCF to determine the ratio in simplest form.	1) $\dfrac{15}{55} =$ 2) $\dfrac{2}{10} =$

2) $\dfrac{6}{15}$

Identify the Greatest Common Factor (GCF) common to both the numerator and denominator. The GCF is 3. Next, divide both the numerator and denominator by the GCF to determine the ratio in simplest form.

$$\frac{6}{15} = \frac{6 \div 3}{15 \div 3} = \frac{2}{5}$$

3) You survey your class and 4 like vanilla ice cream and 14 like chocolate ice cream. Write the ratio of people who like vanilla to people who like chocolate ice cream, then write the ratio in simplest form.

Objective 2 – Multiply and divide by decimal values to introduce percentages

In this section, you will apply percentages to real world examples. This type of math modeling explores relative change and absolute change. In order to accurately calculate and interpret your results, we should briefly discuss percentages in general, then four conversion ideas.

The word "percent" means "per 100" or "out of 100" or "divided by 100". For example, 39% means 39 out of 100. Percent's can also be expressed as fractions or decimals, since they are often used to indicate some part of a whole. So, 39% can also be written as $\frac{39}{100}$ or .39.

There are four conversion types you should be familiar with. They are as follows:

- To convert a percentage to a fraction: The 100 becomes the denominator and do not use the % symbol.
  - Example: $23\% = \frac{23}{100}$ (NOTE: If necessary, reduce the fraction to simplest terms.)
- To convert a percentage to a decimal: Move the decimal point two places to the left and do not use the % symbol.
  - Example: $75\% = 0.75$
- To convert a decimal to a percentage: Move the decimal point to places to the right and use the % symbol.
  - Example: $0.81 = 81\%$
- To convert a fraction to a percentage: Divide the numerator by the denominator to change the fraction to decimal form. Then, move the decimal point two places to the right and use the % symbol.
  - Example: $\frac{4}{16} = 4 \div 16 = 0.25 = 25\%$

Guided Examples	Practice
Convert and calculate the following examples: 1) Convert 19% to a fraction. 100 is the denominator and do not use the % symbol, so the conversion is $\frac{19}{100}$. 2) Convert 46% to a decimal. Move the decimal point two places to the left and do not use the % symbol, so the conversion is 0.46 3) Convert 0.028 to a percentage. Move the decimal point two places to the right and use the % symbol, so the conversion is 28% 4) Convert $\frac{7}{8}$ to a percentage. Divide the numerator by the denominator. The result is $7 \div 8 = 0.875$. Now, move the decimal point two places to the right and use the % symbol, so the final conversion is 87.5%.	Convert and calculate the following examples: 1) Convert 83% to a fraction. 2) Convert 14% to a decimal. 3) Convert 0.57 to a percentage. 4) Convert $\frac{3}{5}$ to a percentage. 5) You decide to purchase an item with an original cost of $250 that is discounted 30%. What is the dollar amount of the discount?

5) You make a purchase of $48 and are required to pay 6% sales tax. What is the dollar amount of the sales tax?

Convert the 6% to a decimal and multiply the two values. The sales tax calculation is $48 \cdot 0.06 = \$2.88$.

6) You make an online purchase of $120 and must pay 8% shipping and handling fees. What is the dollar amount of the S & H fees?

Convert 8% to a decimal and multiply the two values. The S & H calculation is $120 \cdot 0.08 = \$9.60$.

7) Your dinner costs $36 and you want to leave a 20% tip. What is the dollar amount of the tip?

Convert 20% to a decimal and multiply the two values. The tip calculation is $36 \cdot 0.20 = \$7.20$.

6) Your weekly earnings of $600 is taxed at the rate of 15%. What is the dollar amount of your taxes?

7) Because of the concern of global warming, everyone is asked to reduce their carbon footprint. An average U.S. person has a carbon footprint of 20 tons. If you are asked to reduce your carbon footprint by 10%, what is the amount of reduction in tons?

Objective 3 – Percent conversions

In order to analyze the uses and abuses of percentages, you will develop the ability to accurately convert from percent to decimal and decimals to percent. Additionally, your precise understanding of percent conversions will aid in your ability to accurately reason and logic numerical challenges.

Converting from percent to decimal requires division. The percent symbol means "divided by 100". For example, 15% is $\dfrac{15}{100} = 0.15$. The short way to convert from a percent to a decimal is to move the decimal point two places to the left and remove the percent symbol.

Converting from decimal to percent requires multiplication by 100. For example, $0.825 \cdot 100 = 82.5\%$. The short way to convert from decimal to percent is to move the decimal point two places to the right and add the percent symbol.

Guided Examples

Guided Examples	Practice
Convert the following from percent to decimal:	Convert the following from percent to decimal:
1) $8.5\% = 0.085$	1) $7.38\% =$
$8.5\% = \dfrac{8.5}{100} = 0.085$ or move the decimal point two places to the left and remove the percent symbol.	2) $75\% =$
2) $200\% = 2.0$ or 2	3) The price of gas today is 200% of the price last year. Does that mean the price has doubled?
No decimal point is given, so it is placed at the end of the number (after the 0). The decimal point is moved two places to the left and remove the percent symbol.	Convert the following from decimal to percent:
Convert the following from decimal to percent:	4) $0.25 =$
3) $0.068 = 0.068 \cdot 100 = 6.8\%$	5) $0.05 =$
Multiply the decimal by 100 or move the decimal point two places to the right, and add the percent symbol.	6) Can it be true that a 10 year old child weighs 30% more than a 6 year old child?

Objective 4 – Multiply or divide using scientific notation format

The scientific notation format was developed to easily represent very large or very small numbers. This section reviews the conversion of writing very large or very small numbers (standard form) in scientific notation form, especially as it relates to federal spending and the federal deficit.

A number in scientific notation is written as the product of a number (integer or decimal) and a power of 10 in the form $a \times 10^b$ where (for us) a is a rational number greater than 1 and less than 10 and b (the exponent) is an integer. There are three parts to consider when converting from standard form (decimal) to scientific notation:

- the rational number a
- the value of the exponent
- is the exponent positive or negative

The rational number (a) in scientific notation form always has only one non-zero digit to the left of the decimal point. The absolute value of the exponent indicates how many "spaces" you moved your decimal point.

Additionally, if the original number you are working with is greater than 1, the exponent is positive and if the original number you are working with is less than 1, the exponent is negative.

Guided Examples

Practice

Write the following numbers in scientific notation form:	Write the following numbers in scientific notation form:
1) $4,200,000 = 4.2 \times 10^6$ Move the decimal point so there is only one digit to the left of the decimal point (4.2), you moved the decimal point 6 places (exponent 6) and since your beginning number is large (exponent is positive). 2) $3,560.2 = 3.5602 \times 10^3$ 3) $0.00081 = 8.1 \times 10^{-4}$ Move the decimal point so there is only one digit to the left of the decimal point (8.1), you moved the decimal point 4 places (absolute value of the exponent is 4) and since your beginning number is small (less than 1 – the exponent is negative).	1) $315,000 =$ 2) Currently, the federal deficit is approximately $\$17,000,000,000,000$ (17 trillion dollars). Write this number in scientific notation. 3) $0.0037 =$ 4) The diameter of a typical bacterium is 0.000001 meter. Write this number in scientific notation.

To convert from scientific notation to standard form, there is a pattern:

- If the exponent is positive, the exponent designates you move the decimal point to the right by the exponent value.
- If the exponent is negative, the exponent designates you move the decimal point to the left by the exponent value.

Guided Examples

Practice

Write the following numbers in standard form:	Write the following numbers in standard form:
5) $3.8 \times 10^4 = 38,000$ The exponent is positive, so move the decimal point to the right 4 places. The positive exponent indicates a "large" number. 6) $9.23 \times 10^6 = 9,230,000$ The exponent is positive, so move the decimal point to the right 6 places. 7) $4.71 \times 10^{-3} = 0.00471$ The exponent is negative, so move the decimal point 3 places to the left. The negative exponent indicates a "small" number. 8) $8.9 \times 10^{-6} = 0.0000089$ The exponent is negative, so move the decimal point 6 places to the left.	5) $2.9 \times 10^3 =$ 6) The Earth is 1.496×10^8 km from the Sun. Write this number in standard form. 7) $4.3 \times 10^{-2} =$ 8) An average cell has an approximate diameter of 6×10^{-6} meters. Write this number in standard form.

For our federal revenue or deficit calculations involving the population of the United States may require us to combine (multiplying or dividing) large numbers.

If we are multiplying two values in scientific notation form,
- Multiply the rational numbers
- Add the exponents

If we are dividing two values in scientific notation form,
- Divide the rational numbers
- Subtract the exponents

<u>Guided Examples</u>   <u>Practice</u>

Multiply the scientific notation values:	Multiply the scientific notation values:
9) $\left(3.1\times10^5\right)\left(2.5\times10^3\right)=7.75\times10^8$	9) $\left(3.2\times10^6\right)\left(2.5\times10^2\right)=$
Multiply the values: $3.1\cdot2.5=7.75$ Add the exponents: $10^{5+3}=10^8$	
10) $\left(4.1\times10^5\right)\left(2\times10^7\right)=8.2\times10^{12}$	10) $\left(1.6\times10^3\right)\left(6\times10^7\right)=$
Multiply the values: $4.1\cdot2=8.2$ Add the exponents: $10^{5+7}=10^{12}$	
Divide the scientific notation values:	Divide the scientific notation values:
11) $\dfrac{9.3\times10^8}{3.1\times10^2}=3\times10^6$	11) $\dfrac{8.6\times10^{12}}{4.3\times10^3}=$
Divide the values: $9.3\div3.1=3$ Subtract the exponents: $10^{8-2}=10^6$	
12) $\dfrac{8.64\times10^{14}}{3.2\times10^3}=2.7\times10^{11}$	12) $\dfrac{9.2\times10^5}{5\times10^2}=$
Divide the values: $8.64\div3.2=2.7$ Subtract the exponents: $10^{14-3}=10^{11}$	

Objective 5 – Decimal rounding

Understanding the concept of significant digits will better equip you to assess the considerations required in dealing with numerical literacy and calculation challenges. In this regard, there are two areas for us to explore. Place value and rounding are used to describe results with accuracy and precision.

Place value refers to the positioning of either a single digit in a whole number or in a number containing a decimal. The key to determining place value is to become familiar with the below place value map.

Place value chart												
millions	hundred thousands	ten thousands	thousands	hundreds	tens	ones	decimal	tenths	hundredths	thousandths	ten thousandths	

Next, we will use the place value chart to develop an understanding of the process of rounding numbers as it applies to the rounding with significant digits.

The process of rounding numbers is a two-step process as follows:

- Step 1: Decide which place value is most important.
- Step 2: Look at the number in the place to the *right*.
  - o If the value in this next place is less than five $(0 \text{ to } 4)$, there is no change to the place value.
  - o If the value in this next place is greater than or equal to five $(5 \text{ to } 9)$, there is a change of $+1$ to the place value.
  - o All values to the right of the designated place value are no longer used in the rounded answer and zeros are used for place holder values.

Guided Examples

Practice

Round the following numbers to the value indicated	Round the following numbers to the value indicated
1) 2643.7 to the nearest hundred is 2600 . The hundred place value is designated. This is the value 6 . Look at the number in the place to the right of the 6 . This is the value 4 . Since 4 is less than 5 , then the value 6 is not changed. The 43.7 is not used in the rounded answer (replaced with zeros).	1) 428.1 to the nearest ten. 2) 526,034 to the nearest ten thousand. 3) 9833.5 to the nearest hundred.

2) 37.923 to the nearest hundredth is 37.92 .

The hundredth place value is designated. This is the value 2 . Look at the number in the place to the right of the 2 . This is the value 3 . Since 3 is less than 5 , then the value 2 is not changed. The 3 is not used in the rounded answer.

3) 14,692.7 to the neared thousand is 15,000 .

The thousand place value is designated. This is the value 4 . Look at the number in the place to the right of the 4 . This is the value 6 . Since 6 is greater than 5 , then the value 4 is changed to 5 . The 692.7 is not used in the rounded answer (replaced with zeros).

4) 8.62 to the nearest tenth is 8.6 .

The tenth place value is designated. This is the value 6 . Look at the number in the place to the right of the 6 . This is the value 2 . Since 2 is less than 5 , then the value 6 is not changed. The 2 is not used in the rounded answer.

4) 8.2137 to the nearest thousandth

5) 439.25 to the nearest tenth

6) 45.338 to the nearest hundredth

Chapter 4 Managing Money

Learning Objectives
Objective 1 – Use order of operations for real numbers Objective 2 – Solve proportions Objective 3 – Find the amount, base, or percent in a percent problem by solving an equation Objective 4 – Solve linear equations using addition and multiplication Objective 5 – Evaluate radical expressions

Objective 1 – Use order of operations for real numbers

PEMDAS Review

Taking control of your finances requires you to keep track of monthly expenses. To annualize (12 months) these expenses, you multiply by 12. Some expenses are semiannual (twice a year), so you may add expenses then multiply by 2. Adjusting cash flow or a budget may require you to work with negative numbers. This review section explores all these rules so you are able to confidently make personal finance calculations in this section.

You will frequently use the PEMDAS acronym to help guide your personal finance decision making when evaluating algebraic expressions:

Parentheses | Exponents | Multiplication | Division | Addition | Subtraction

1. Perform the operations inside a parentheses first
2. Then exponents
3. Then multiplication and division, from left to right
4. Then addition and subtraction, from left to right

Integer Review

1. **Adding Rules with examples:**

Positive + Positive = Positive: $4 + 7 = 11$
Negative + Negative = Negative: $(-5) + (-2) = -7$

Sum of a negative and a positive number: You need to subtract the absolute values of the two numbers then use the sign of the number with the larger absolute value.

$(-11) + 2 = -9$
$4 + (-5) = -1$
$(-2) + 9 = 7$
$8 + (-5) = 3$

2. **Subtracting Rules with examples:**

Negative - Positive = Negative: $(-7)-6=-7+(-6)=-13$

Positive - Negative = Positive + Positive = Positive: $7-(-2)=7+2=9$

Negative - Negative = Negative + Positive = You need to subtract the absolute values of the two numbers then use the sign of the number with the larger absolute value. (*NOTE: Two negatives next to each other become positive*)

$(-2)-(-8)=(-2)+8=6$

$(-3)-(-7)=(-3)+7=4$

3. **Multiplying Rules with examples:**

Positive $\times$ Positive = Positive: $8\cdot2=16$

Negative $\times$ Negative = Positive: $(-3)\cdot(-7)=21$

Negative $\times$ Positive = Negative: $(-8)\cdot5=-40$

Positive $\times$ Negative = Negative: $5\cdot(-3)=-15$

4. **Dividing Rules with examples:**

Positive $\div$ Positive = Positive: $18\div3=6$

Negative $\div$ Negative = Positive: $(-36)\div(-4)=9$

Negative $\div$ Positive = Negative: $(-24)\div3=-8$

Positive $\div$ Negative = Negative: $15\div(-3)=-5$

Guided Examples	Practice

Guided Examples:

1) Evaluate the expression $5\cdot(3+6)$

Because of the order of operations, you will perform operations inside the parentheses first, $3+6=9$. Next, you multiply $5\cdot9$. The value of the expression is 45.

2) Calculate the total when the following college expenses are paid twice a year: Tuition of 3000, 800 in student fees and 350 for textbooks.

The expression can be set up as $2(3,000+800+350)$. Because of the order of operations, you will perform operations inside the parentheses first.

So, $2(4,150)$ totals $8,300$.

Practice:

1) $(8+12)\cdot5$

2) $5+2(3+6)^2$

3) You spend 20 a week for coffee and 180 per month for groceries. If a month is four weeks long, calculate your total coffee and grocery expense?

4) You decide to lease/buy a car with the following conditions: a one-time fee of 300, a down payment of 2000 and monthly payments of 210. What is the annual (12 month cost) of this plan?

3) Determine if you have a positive or negative monthly cash flow with the following details:

Income: Part time job $\$400$ per month, scholarship $\$300$ per month and a student loan of $\$250$ per month.

Expenses: Rent $\$725$ per month, groceries $\$65$ per month, phone $\$60$ per month and entertainment is $\$180$ per month.

The expression can be set up as $\left(400+300+250\right)-\left(725+65+60+180\right)$. This becomes $950-1030=-80$. This represents a negative cash flow of $\$80$ per month.

5) $15+3=$

6) $20-\left(-5\right)=$

7) $-9+\left(-2\right)=$

8) $-6\cdot 10=$

9) $30\div -3=$

10) $9+\left(-4\right)=$

11) $-5\cdot -2=$

12) $-20\div -2=$

13) $-18+12=$

Objective 2 – Solve Proportions

A proportion is an equation which states that two ratios (fractions) are equal to each other. If one term of a proportion is not known (designated by the variable x), the concept of cross multiplication can be used to find the value of the unknown term.

The numerator is the top number in a fraction and the denominator is the bottom number in a fraction. For example, in the fraction $\frac{3}{8}$, 3 is the numerator and 8 is the denominator.

To cross multiply, take the first fraction's numerator and multiply it by the second fraction's denominator. Then take the first fraction's denominator and multiply it by the second fraction's numerator.

When the terms of a proportion are cross multiplied, the terms are set these equal to each other and the fraction form is no longer used.

Guided Examples	Practice
1) Solve the proportion for: $\frac{4}{12} = \frac{9}{x}$ To set up the cross multiplication, $4 \cdot x = 12 \cdot 9$. Multiply the terms on each side becomes $4x = 108$. To isolate the variable and solve for x, divide both sides of the equation by 4 and you get $x = 27$. 2) Solve the proportion for x: $\frac{3}{8} = \frac{6}{x+4}$ To set up the cross multiplication, make sure the 3 is distributed to all terms in the denominator, so $3(x+4) = 8 \cdot 6$. Next, distribute the 3 to all terms in the parentheses by multiplication. When you distribute, the parentheses are no longer required. You get $3x + 12 = 48$ To isolate the variable, you do the opposite operation of add 12, which is to subtract 12 from both sides of the equation. The result is $3x = 36$. Finally, you do the opposite operation of multiply by 3, which is to divide both sides by 3. The result is $x = 12$.	1) Solve the proportion $\frac{9}{36} = \frac{4}{x}$ 2) Solve the proportion $\frac{15}{x} = \frac{20}{12}$ 3) Solve the proportion $\frac{2x+3}{6} = \frac{5}{2}$ HINT: Make sure you multiply the 2 to both terms in the parentheses $2(2x+3)$. 4) Set up a proportion to solve. There are 12 students who wear glasses in your class of 20. If you go to a class with 50 students, how many will be wearing glasses? HINT: Be sure to "line up" the units in both your ratios: $\frac{\text{students wear glasses}}{\text{students in class}} = \frac{\text{students wear glasses}}{\text{students in class}}$

Objective 3 – Find the amount, base, or percent in a percent problem by solving an equation

Percentages refer to fractions of a whole. In our application of earning a percentage of interest on money invested, the percentage refers to earning (or receiving) a fraction of the money you invested in the form of interest earned the bank will pay to you.

To multiply a value by a percent requires that the percent be converted to a decimal so you can multiply the values. For example, to complete the operation $2.6\% \cdot \$250$ requires the percentage be converted to a decimal before multiplying.

To convert from percentage to decimal, **divide the decimal by 100** AND remove the "%" sign OR the quick way to divide the decimal by 100 is to **move the decimal point 2 places to the left** and remove the "%" sign.

To convert from decimal to percentage, **multiply the decimal by 100** AND write the "%" sign OR the quick way to multiply by 100 is to **move the decimal point 2 places to the right** and write the "%" sign.

Guided Examples

Practice

Guided Examples	Practice
1) 3.9% converts to the decimal 0.039	1) Convert 200% to a decimal
2) 7% converts to the decimal 0.07	2) Convert 6.1% to a decimal
3) the decimal 0.45 converts to 45%	3) Convert 8% to a decimal
4) the decimal 0.071 converts to 7.1%	4) Convert 0.079 to a percentage
5) the decimal 23 converts to 2300%	5) Convert 4.1 to a percentage
6) What is 8.2% of 60?	6) What is 5.4% of 230?
First, convert the percent to a decimal 8.2% becomes 0.082. Next, multiply $0.082 \cdot 60 = 4.92$.	
	7) What is 38% of 750?
	8) You deposit $500 in a bank or credit union account with an annual (simple) interest rate of 4%. How much interest do you earn after one year? For now, multiply the deposit and the interest rate (HINT: First convert the interest percent to a decimal).

Objective 4 – Solve linear equations using addition and multiplication

Solving one step equations is a required calculation for working with loan basics. Your calculations will involve the principal (amount of money you borrow), the interest you are charged and your efforts to gradually pay down the principal.

Solving these linear equations is the task of isolating (solving for) the variable x. To isolate the variable on one side of the equal sign, you must keep in mind the following:

- start on the side with the variable
- look to do the opposite operation – always start with addition or subtraction
- what you do to one side of the equal sign, you must do to the other side of the equal sign.
- then look to do the opposite operation – multiplication or division to finish

Guided Examples

Practice

| Solve the linear equation.

1) $5x - 1 = 19$

$\quad 5x - 1 = 19$ start on the left side(side with x)
$\quad \underline{+1 \quad +1}$ the opposite of -1 is $+1$ and do
$\qquad$ this to both sides

$\quad \dfrac{5x}{5} = \dfrac{20}{5}$ the opposite of multiply by 5
$\qquad$ is to divide by 5 and do this to
$\qquad$ both sides
$\quad x = 4$

2) $3x + 5 = 35$

$\quad 3x + 5 = 35$ start on the left side(side with x)
$\quad \underline{-5 \quad -5}$ the opposite of $+5$ is -5 and do
$\qquad$ this to both sides
$\quad \dfrac{3x}{3} = \dfrac{30}{3}$ the opposite of multiply by 3
$\qquad$ is to divide by 3 and do this to
$\qquad$ both sides. The result is
$\quad x = 10$ | Use algebraic concepts to solve the following linear equation:

1) $4x + 2 = 26$

2) $5x - 12 = 18$

3) $3x - 15 = 12$ |

Name:_____Date:_____

Instructor:_____Section:_____

<u>Objective 5 – Evaluate radical expressions</u>

Using savings plan formulas, investment gains and rates or return require calculations using an understanding of roots and the radical symbol.

The $\sqrt{}$ symbol is called the radical symbol. A square (second) root is written as $\sqrt[2]{}$. In this case, the index number 2 is not written, so $\sqrt[2]{}$ is the same as $\sqrt{}$. There are other roots to consider. A cube (third) root is written as $\sqrt[3]{}$, a fourth root is written as $\sqrt[4]{}$ and a fifth root is written as $\sqrt[5]{}$.

When you use the radical symbol, you have to ask yourself, "What same number times itself by the index number is equal to the value under the radical sign?

Let's do a couple of examples to become familiar with this notation and how to use it.

EX1: $\sqrt[2]{36}$. What two same numbers multiply to 36? We state the square root of $36 = 6$, because $6 \cdot 6 = 36$
EX2: $\sqrt[4]{81}$. What four same numbers multiply to 81? We state the fourth root of $81 = 3$, because $3 \cdot 3 \cdot 3 \cdot 3 = 81$.

Now, most values are not perfect squares. When this happens, your answer will be a decimal approximation. For these approximations, a calculator can be a useful utility.

EX3: $\sqrt{18} \approx 4.24$ (approximate). This means that $4.24 \cdot 4.24$ is about 18.

<u>Guided Examples</u>

Evaluate the radical expression:

1) $\sqrt{9}$ What two same numbers multiply to 9? The square root of $9 = 3$.

2) $\sqrt[3]{125}$ = What three same numbers multiply to 125? The cube root of $125 = 5$.

3) $\sqrt[4]{256}$ What four same numbers multiply to 256? The fourth root of $256 = 4$.

4) $\sqrt[5]{32}$ What five same numbers multiply to 32? The fifth root of $32 = 2$.

<u>Practice</u>

Evaluate the radical expression:

1) $\sqrt{144} =$

2) $\sqrt[3]{64} =$

Approximate the radical expression:

3) $\sqrt{24} =$

4) If you invest $\$3,000$ that grows to $\$8,400$ over four years, what is the percent increase? The radical expression looks like this: $\sqrt[4]{\dfrac{8400}{3000}} =$

Chapter 5 Statistical Reasoning

Learning Objectives
Objective 1 – Read and interpret graphs

Objective 1 – Read and Interpret Graphs

A **bar graph** is a two dimensional visual diagram in which the numerical counts of a categorical variable are represented by the height (or length) of rectangles of equal width. This visual display is used to compare characteristics of two types of data groups.

There are several items to consider when reading, analyzing and interpreting a bar graph. These are the graph title, the axes labels, scale and the height of the bars. This attention to graph details will improve the accuracy of your interpretation of the data presented in the graph.

First, locate the horizontal line on the graph (*x*-axis) that runs along the bottom of the bar graph. Typically, this is the location of information describing the data group for each bar.

Next, locate the vertical line on the left side of the bar graph (*y*-axis). This is the location of information describing the numerical value of the data group. The scale of this numerical data (hundreds, millions) is critical to your interpretation and analysis and is almost limitless in how it is presented.

Finally, your ability to read and interpret this correlating information is instrumental in your analysis. In your analysis of the data presented in the bar graph, you will correlate (or match) the data on the horizontal line (*x*-axis) with the data on the vertical line (*y*-axis).

Guided Example   Practice

Analyze the below bar graph:

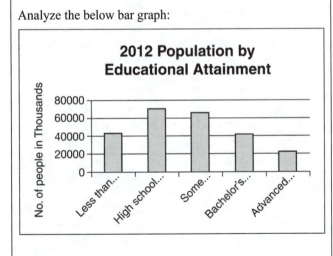

1) Which level of education has the highest population? High school graduate

2) Which two levels of education have approximately the same level of education? Less than HS and Bachelor's

3) What is the approximate number (in thousands) of people with advanced degrees? 21,000 (in thousands)

Analyze the below bar graph:

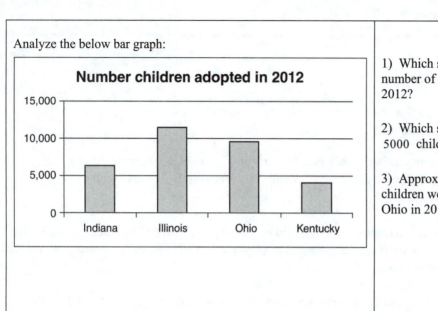

1) Which state had the highest number of children adopted in 2012?

2) Which state had fewer than 5000 children adopted in 2012?

3) Approximately how many children were adopted in the state of Ohio in 2012?

Chapter 6 Putting Statistics to work

Learning Objectives
Objective 1 – Calculate average Objective 2 – Find the square root of a number Objective 3 – Find the percent by solving an equation Objective 4 – Evaluate and approximate radical expressions

Objective 1 – Calculate average

A measure of central tendency is the calculation of a value that attempts to numerically describe or identify the central position within a set of data. You may be familiar with the mean (also called the average) as a measure of central tendency. Additionally, we will explore the median and the mode.

The mean, median and mode are all measures of central tendency, but they are calculated differently and under varied circumstances, some measures of central tendency are more appropriate and accurate to use in characterizing, analyzing and interpreting your data.

Measure of central tendency	Description
Mean	Also called **average value**. The mean is calculated by adding up all the scores and dividing the total by the number of data items in the illustration.
Median	Is the **middle value**. The median represents the value in the exact middle of a number of data items when the data are arranged in ascending (lowest to highest) order or descending (highest to lowest) order. In this case, half of the data items are above the median and half are below the median. NOTE: If the data items total to an even number, then the median is the average of the two middle terms when they are written in order.
Mode	Is the **most common value**. The mode is identified as the value with the highest frequency of all data items in the illustration. To clarify, it is the most common value or the value that appears most often.

Guided Examples

Calculate the measures of central tendency – mean, median and mode for the data provided:

1) A study of newborn birth weights at the local hospital on a specific day showed that 7 babies were born and their weights (in pounds) were: $7, 5, 6, 9, 12, 6, 11$.

$$\text{Mean} = \frac{7+5+6+9+12+6+11}{7} = \frac{56}{7} = 8 \text{ pounds}$$

Practice

Calculate the measures of central tendency – mean, median and mode for the data provided:

1) A study of the age of children (in years) in a small daycare show the ages to be: $2, 5, 7, 4, 2, 2, 6$.

a) Mean =

For median, the original sequence becomes $5,6,6,7,9,11,12$ when the values are arranged in an ascending numerical sequence. Since 7 is the middle term of this sequence, the median is 7 pounds.

NOTE: If the data items are $6,6,7,9,11,12$, the data items total an even number, so you must average the two middle terms. With these data items, the median is $\dfrac{7+9}{2} = \dfrac{16}{2} = 8$.

Mode $= 6$ pounds. The most common value is 6.

b) Median =

c) Mode =

Objective 2 – Find the square root of a number

Often times, the collection of data can be described as clustering closely together or far apart. When you get a cup of your favorite coffee from your local coffee shop, is it consistently the same delicious taste or does the taste vary greatly from visit to visit? Your response is known as a measure of variation.

The measure of variation describes how closely grouped or scattered a collection of data can be. The three measures of variation are the range, the variance and the standard deviation. To accurately work with these measures of variation requires us to correctly find the square root of a number.

The $\sqrt{}$ symbol is called the radical symbol. A square (second) root is written as $\sqrt[2]{}$. In this case, the index number 2 is not written, so $\sqrt[2]{}$ is written as $\sqrt{}$. When you use the radical symbol, you have to ask yourself, "What same number times itself, the index number of times, is equal to the value of the radicand (the value under the radical sign)?

Let's do an example to become familiar with this notation and how to use it.

EX1: $\sqrt[2]{25} = \sqrt{25}$. What two same numbers multiply to 25? We state the square root of $25 = 5$, because $5 \cdot 5 = 25$.

NOTE: In this section, we are only working with square roots, so your precise answer or approximation should be a non-negative (a positive) value.

<u>Guided Examples</u>

Calculate the square root of the following examples:
1) $\sqrt{16} = 4$ Because $4 \cdot 4 = 16$.
2) $\sqrt{144} = 12$ Because $12 \cdot 12 = 144$.
3) $\sqrt{28} \approx 5.29$ The $\sqrt{28}$ will be an approximation because there are no two exact numbers that multiply to 28 (28 is not a perfect square). A calculator is helpful when approximating square roots. To check, $5.29 \cdot 5.29 \approx 28$.

<u>Practice</u>

Calculate the square root of the following examples:
1) $\sqrt{9} =$
2) $\sqrt{64} =$
3) Approximate the $\sqrt{48} =$

Objective 3 – Find the percent by solving an equation

In this section, you will be challenged to calculate and interpret numerical perspective across a variety of math modeling applications. In order to accurately calculate, analyze and translate your results, we should briefly discuss numerical perspectives using a general format.

The general form builds a proportion using two ratios set equal to each other, then solve by cross multiplication. A proportion is a label given to a statement that two ratios are equal. It can be written as:

$\frac{a}{b} = \frac{c}{d}$. When two ratios are equal, then the cross products of the ratios are equal. For this general form, that

means that $a \cdot d = b \cdot c$. The numerator of the first fraction is multiplied by the denominator of the second fraction and set equal to the denominator of the first fraction multiplied by the numerator of the second fraction.

Specifically, for math modeling and to advance numerical perspective, a specific proportion is created as follows:

- $\frac{is}{of} = \frac{\%}{100}$.

We can use this format to answer the question, "What is 30% of 60 ?" The challenge will be to put the values in the correct location prior to your calculation and solve for the unknown (x). We always have to work with three values:

1) The 100 never changes location, the 30 is linked with the % symbol and the 60 is linked with the word *of*.

The result is $\frac{x}{60} = \frac{30}{100}$. We now have to solve for x. Use cross products to get $100x = 60 \cdot 30$. So, $100x = 1,800$.

Lastly, divide both sides by 100 to get $x = 18$.

Guided Examples	Practice
Solve the math modeling problems for x. 1) What is 45% of 120 ? $\frac{x}{120} = \frac{45}{100}$ Next, cross multiplication results in $100x = 120 \cdot 45$. By multiplying the right side, you get $100x = 5400$. Last, divide both sides by 100, to get $x = 54$. What three values are we working with? The 100 never changes location, the 45 is linked with the % symbol and 120 is linked with the word *of*. 2) 30 is what percent of 150 ? $\frac{30}{150} = \frac{x}{100}$ Next, $30 \cdot 100 = 150 \cdot x$. $3,000 = 150 \cdot x$. $20 = x$. The answer is $x = 20\%$. What three values are we working with? The 100 never changes location, the 30 is linked with the word *is* and the 150 is linked with the word *of*.	Solve the math modeling problems for x. 1) What is 20% of 300? 2) 60 is what percent of 200 ? 3) 40% of what number is 50 ?

Objective 4 – Evaluate and approximate radical expressions

When you analyze a statistical study, you will be expected to determine the statistical significance of the numerical data collected. To precisely calculate and quantify statistical significance requires you to evaluate and approximate radical expressions.

Specifically, the radical symbol $\left(\sqrt{}\right)$ is used when evaluating margin of error and the confidence interval. The mathematical relationship is as follows:

$$\textbf{margin of error} \approx \frac{1}{\sqrt{n}}$$

Guided Examples	Practice
Approximate the margin of error using the formula:	Approximate the margin of error using the formula:
1) $\dfrac{1}{\sqrt{250}} \approx 0.063$	1) $\dfrac{1}{\sqrt{300}} \approx$
To calculate an approximate answer will require the use of calculator. The formula means $1 \div \sqrt{250}$ and can be entered into a calculator. The result is $1 \div \sqrt{250} = 1 \div 15.81 \approx 0.063$.	
2) $\dfrac{1}{\sqrt{500}} \approx .044$	2) $\dfrac{1}{\sqrt{50}} \approx$
To calculate an approximate answer will require the use of calculator. The formula means $1 \div \sqrt{500}$ and can be entered into a calculator. The result is $1 \div \sqrt{500} = 1 \div 22.36 \approx .045$.	3) Calculate the margin of error for a survey of 200 people $(n = 200)$. Use the formula **margin of error** $\approx \dfrac{1}{\sqrt{n}}$.
3) Calculate the margin of error for a survey of 600 people $(n = 600)$. Use the formula **margin of error** $\approx \dfrac{1}{\sqrt{n}}$. When you substitute 600 for *n* in the formula, the formula means $1 \div \sqrt{600}$ and can be entered into a calculator. The result is $1 \div \sqrt{600} = 1 \div 24.49 \approx .041$.	

Name:_____ Date:_____

Instructor:_____ Section:_____

Chapter 7 Probability: Living with the Odds

Learning Objectives

Objective 1 – Evaluate an exponential expression
Objective 2 – Multiply real numbers
Objective 3 – Write or evaluate algebraic expressions using exponents

Objective 1 – Evaluate an exponential expression

To develop an accurate understanding of the fundamentals of probability, we will acquire the ability to evaluate an exponential expression. Specifically, we will advance our understanding, interpretation and calculation of the Fundamental Counting Principle and combining probabilities.

These concepts require us to further explore and precisely apply exponents. In the general form, a^n, a is the base number and n is the exponent. An exponent is a number that tells how many times the base number is multiplied (a factor).

For example, 5^3 indicates that the base number 5 is multiplied 3 times. To determine the value of 5^3, multiply $5 \cdot 5 \cdot 5$ which equals 125 $(5 \cdot 5 \cdot 5 = 125)$.

When combining probabilities, fractions are frequently used. The use of fractions in an exponential expression is similar to the previous example.

For example, $\left(\dfrac{3}{4}\right)^2$ indicates that the numerator (3) AND the denominator (4) are multiplied 2 times. To

continue, the value of $\left(\dfrac{3}{4}\right)^2 = \dfrac{3 \cdot 3}{4 \cdot 4} = \dfrac{9}{16}$.

If you are multiplying fractions that do not use an exponent, your calculation process is similar – multiply across the numerator and multiply across the denominator. A common denominator is not required when you multiply

fractions. If necessary, simplify your answer. For example, $\left(\dfrac{3}{5}\right) \cdot \left(\dfrac{2}{4}\right) \cdot \left(\dfrac{1}{3}\right) = \dfrac{3 \cdot 2 \cdot 1}{5 \cdot 4 \cdot 3} = \dfrac{6}{60}$. This further reduces by

finding and dividing by the GCF (Greatest Common Factor), so the final answer looks like $\dfrac{6}{60} = \dfrac{6 \div 6}{60 \div 6} = \dfrac{1}{10}$.

Guided Examples	Practice
Calculate the following exponential expressions: 1) $4 \cdot 2 \cdot 3 = 24$ Consider the order of operations with this problem. Since the operations are all multiplication, just multiply the numbers from left to right to get 24.	Calculate the following exponential expressions: 1) $4 \cdot 2 \cdot 3 \cdot 3 =$

2) You are eating at a cafeteria that offers 3 types of salads, 2 types of meats, 5 different vegetables and 4 different desserts. How many ways can you choose a different meal?

$3 \cdot 2 \cdot 5 \cdot 4 = 120$ different ways

3) $\left(\dfrac{4}{5}\right)^3 =$

To solve, remember the exponent (3), is distributed to both the numerator (4) and denominator (5). The result is $\left(\dfrac{4}{5}\right)^3 = \dfrac{4 \cdot 4 \cdot 4}{5 \cdot 5 \cdot 5} = \dfrac{64}{125}$

4) $\left(\dfrac{6}{5}\right) \cdot \left(\dfrac{5}{4}\right) \cdot \left(\dfrac{4}{3}\right) =$

Multiply straight across the numerator and straight across the denominator (no common denominator is required when you multiply fractions).

$\left(\dfrac{6}{5}\right) \cdot \left(\dfrac{5}{4}\right) \cdot \left(\dfrac{4}{3}\right) = \dfrac{6 \cdot 5 \cdot 4}{5 \cdot 4 \cdot 3} = \dfrac{120}{60} = 2$

2) You look in your closet and have 4 tops, 5 pants and 12 pair of shoes. How many different outfits can you make (assuming all items match)?

3) $\left(\dfrac{5}{6}\right)^2 =$

4) $\left(\dfrac{5}{10}\right) \cdot \left(\dfrac{4}{9}\right) \cdot \left(\dfrac{3}{8}\right) =$

Objective 2 – Multiply real numbers

When working with the Law of Large Numbers, as it relates to probability, you will explore the concept of expected value. Expected value is a value you would "expect" to achieve if you could repeat a process an "infinite" number of times (Law of Large Numbers).

This concept leads to decision theory that is most often used for business decision making. The expected value model is used in a situation where monetary gains and losses will occur. After calculating the expected value, the optimal choice would be the alternative that makes the most money (has the highest expected value).

The expected value formula requires the multiplication of real numbers. In words, to compute the expected value you would multiply the monetary payoff for each alternative outcome by the probability of the event occurring. The formula is:

$$E(v) = (\text{outcome} \#1 \cdot \text{probability}) + (\text{outcome} \#2 \cdot \text{probability}) + (\text{outcome} \#3 \cdot \text{probability})...$$

Guided Examples   Practice

Guided Examples	Practice
Multiply the real numbers:	Multiply the real numbers:

Multiply the real numbers:

1) $5 \cdot \dfrac{3}{4} =$

There are two ways to multiply these numbers. Choose the method that works best for you:

a) Change the first number to a fraction (denominator $=1$), then multiply the numerators and the denominators. This becomes
$$\frac{5}{1} \cdot \frac{3}{4} = \frac{5 \cdot 3}{1 \cdot 4} = \frac{15}{4} = 3.75$$

b) Use order of operations (PEMDAS) and calculate the problem left to right. The whole number (5) is multiplied to the numerator (3) of the fraction, then the results is divided by the denominator (4) of the fraction. This becomes
$$5 \cdot 3 \div 4 = 3.75$$

2) $\left(12 \cdot \dfrac{2}{3}\right) + \left(15 \cdot \dfrac{1}{3}\right) =$

a) Convert the first number to a fraction (denominator $=1$), then multiply across the numerator and across the denominator.
$$\left(\frac{12}{1} \cdot \frac{2}{3}\right) + \left(\frac{15}{1} \cdot \frac{1}{3}\right) = \left(\frac{12 \cdot 2}{1 \cdot 3}\right) + \left(\frac{15 \cdot 1}{1 \cdot 3}\right) = \left(\frac{24}{3}\right) + \left(\frac{15}{3}\right) = 8 + 5 = 13$$

b) Using the order of operations (PEMDAS), work in the first parentheses first, then the second set of parentheses. Combine the resulting two numbers.
$$(12 \cdot 2 \div 3) + (15 \cdot 1 \div 3) = (24 \div 3) + (15 \div 3) = 8 + 5 = 13$$

Practice — Multiply the real numbers:

1) $8 \cdot \dfrac{4}{5} =$

2) $\left(14 \cdot \dfrac{2}{5}\right) + \left(10 \cdot \dfrac{3}{5}\right) =$

3) You expect your grandmother will send you money for your birthday. Based on past experience, you figure there is a $\frac{3}{10}$ chance your grandmother will send you $20 and a $\frac{7}{10}$ chance she will send you $50. What is your expected value for this event? The formula looks like:

$$E(v) = \left(20 \cdot \frac{3}{10}\right) + \left(50 \cdot \frac{7}{10}\right)$$

a) $E(v) = \left(\frac{20}{1} \cdot \frac{3}{10}\right) + \left(\frac{50}{1} \cdot \frac{7}{10}\right) = \left(\frac{60}{10}\right) + \left(\frac{350}{10}\right) = 6 + 35 = \$41$

b) $E(v) = (20 \cdot 3 \div 10) + (50 \cdot 7 \div 10) = 6 + 35 = \$41$

This means that if your grandmother did this for many, many years – on average you would receive $41 on your birthday.

3) Your community center is giving away raffle cards (no charge to you) and every card is a winner. There is a $\frac{9}{10}$ chance you will win $2 and a $\frac{1}{10}$ chance you will win $10. If you were to receive a large, large quantity of these tickets, what is your expected value for this event?

Objective 3 – Write or evaluate algebraic expressions using exponents

When exploring and studying probability concepts, there are some basic counting techniques employed to discover possible outcomes for a specific situation. In this section, we try to give precise mathematical meaning to questions such as, "How many ways…?, "What are the chances…? and "What is the likelihood…?".

Specifically, these questions make take the form:

- How many ways can you create a 4 digit ATM pin number?
- What are the chances of selecting an all women committee if the group consists of men and women?
- What is the likelihood of rolling a pair of $6's$ in a dice game?

To answer these counting questions, we will explore the concepts of evaluating mathematical expressions using exponents and factorial notation.

First, evaluating mathematical expressions using exponents is helpful when studying products where the same factor may occur more than once. For example, write the following expression using exponents:

- $15 \cdot 15 \cdot 15 \cdot 4 \cdot 4 \cdot 4 \cdot 4 \cdot 4 = (15)^3 \cdot (4)^5$

Please note that when working with exponents, the operation is multiplication. Then, count the number of times a factor repeats – the factor 15 repeats 3 times (3 is the exponent) and the factor 4 repeats 5 times (5 is the exponent).

Next, the use of factorials is a help when exploring and mastering the concepts of permutations and combinations. There is a specialized and specific symbol and pattern associated with factorials.

- Factorial symbol is !. The ! is a shorthand way to denote the multiplication of consecutive positive integers in descending order.
- The pattern is demonstrated as $4! = 4 \cdot 3 \cdot 2 \cdot 1 = 24$
  Again, the pattern is demonstrated as $6! = 6 \cdot 5 \cdot 4 \cdot 3 \cdot 2 \cdot 1 = 720$

NOTE: As you have noticed, the calculation is straightforward and the numbers get large quickly. There is a factorial button (!) on your calculator to help you with large calculations.

Guided Examples	Practice

Write the following expressions using exponents:	Write the following expressions using exponents:
1) $5 \cdot 5 \cdot 5 \cdot 8 \cdot 8 \cdot 8 \cdot 8 =$ Is the operation multiplication? If so, you can use exponents to rewrite this expression. Then, count the number of times a factor repeats – the factor 5 repeats 3 times (3 becomes the exponent) and the factor 8 repeats 4 times (4 becomes the exponent). $5 \cdot 5 \cdot 5 \cdot 8 \cdot 8 \cdot 8 \cdot 8 = (5)^3 \cdot (8)^4$	1) $2 \cdot 2 \cdot 2 \cdot 2 \cdot 7 \cdot 7 \cdot 7 =$

2) $(3)(3)(12)(12)(12)(12)(12) =$

The parentheses indicate multiplication and do not affect your decision.

$(3)(3)(12)(12)(12)(12)(12) = (3)^2 \cdot (12)^5$

Calculate the value of the expressions involving factorials.

3) $5! =$
$5! = 5 \cdot 4 \cdot 3 \cdot 2 \cdot 1 = 120$

4) $(12 - 9)! =$
Recall order of operations (PEMDAS) and work in the parentheses first. So, $(12 - 9)! = (3)! = 3 \cdot 2 \cdot 1 = 6$.

5) $5! - 4! =$
Calculate each factorial value first, then subtract.
So, $5! - 4! = 120 - 24 = 96$.

6) $\dfrac{6!}{3!} =$
Calculate each factorial value first, then divide. So,
$\dfrac{6!}{3!} = \dfrac{720}{6} = 120$.

7) You own a 4 digit bike lock that can use the single digits $0,1,2,3,4,5,6,7,8,9$ (10 numbers). How many ways can you create a different bike lock combination (if you can repeat numbers)?

You have 4 spaces and you can place any one of the 10 numbers in each space, so
$10 \cdot 10 \cdot 10 \cdot 10 = (10)^4 = 10,000$ ways.

2) $(9)(9)(9)(4)(4) =$

Calculate the value of the expressions involving factorials.

3) $4! =$

4) $(10 - 5)! =$

5) $5! - 3! =$

6) $\dfrac{6!}{4!} =$

7) You open a new bank account and you want to set up your new 4 digit ATM pin number. Bank regulations state that you have to use the single digit numbers $0,1,2,3,4,5,6,7,8,9$. How many ways can you create a different ATM pin number (if you can repeat numbers)?

Chapter 8 Exponential Astonishment

Learning Objectives

Objective 1 – Graph exponential functions
Objective 2 – Solve exponential equations
Objective 3 – Evaluate logarithms

Objective 1 – Graph exponential functions

There are two basic growth patterns – linear growth and exponential growth. Linear activity is very common, but there are events that cannot be explained with linear math modeling. For example:

- In biology, microorganisms grow at an exponential rate until the nutrients are consumed.
- In physics, nuclear power activity is best explained with exponential rates.
- In finance, compound interest behaves in an exponential pattern.
- In computer technology, the processing power of computers (Moore's Law) can be described using an exponential pattern.
- In medicine, the way medicine and caffeine in your system wears off is best described by exponential decay.

Using your personal finances, if you receive a $500 wage increase each year, then this is an example of linear growth. If you receive a 10% wage increase each year then this is an example of exponential growth. Decay or decrease works in much the same way for both linear and exponential math modeling.

Our challenge this section will be to examine and interpret linear growth/decay and exponential growth/decay as it relates to graphs and real world applications.

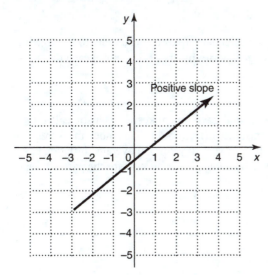
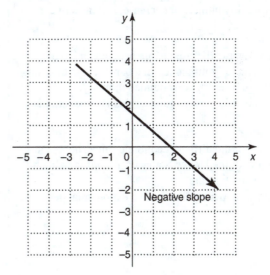

Name:_____Date:_____

Instructor:_____Section:_____

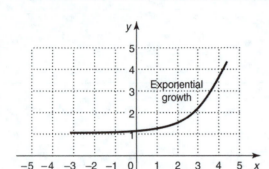

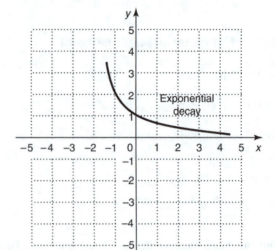

Guided Examples

Practice

Determine if the written example or graph is an example of linear growth, linear decay, exponential growth or exponential decay:

1) The student enrollment at your College or University has increased by 350 people each year for the past 6 years.
This is a yearly numerical increase, so this is linear growth.

2) You have started a healthy living program. You are starting to see the benefits and have lost 2 pounds a week for the past 5 weeks.
This is a weekly numerical decrease, so this is linear decay.

3) You are a fiscally responsible person and have invested some of your money in a bank account earning a 2% rate of interest, compounded monthly.
This is a monthly percent gain, so this is exponential growth.

4) Because of the difficult economic times, homes in certain parts of the United States decreased in value by 7% a year for 6 years.
This is a yearly percent decline, so exponential decay.

Determine if the written example or graph is an example of linear growth, linear decay, exponential growth or exponential decay:

1) You have noticed that around the holidays, the price of gasoline increases $.10 per day for many days.

2) The value of your car decreases by 10% each year for many years.

3) As the calendar moves from summer to fall to winter, the days get shorter by about 3 minutes each day.

4) The federal government keeps a statistic called the CPI (Consumer Price Index). This index measures inflation from year to year. On average, inflation has increased by about 2% each year for the past 10 years.

Objective 2 – Solve exponential equations

You have discovered that exponential growth and decay behaves differently than linear growth and decay. This difference occurs algebraically as well:

- Linear behavior can be expressed as $y = 2x$.
- Exponential behavior can be expressed as $y = 2^x$, where the variable x is an exponent.

This leads to a definition where an exponential equation is one in which a variable occurs as an exponent. We know from experience that to solve an algebraic equation for x requires you have to get the variable on one side of the equation by applying opposite operations to both sides of the equation (the opposite of multiplication is division and the opposite of addition is subtraction).

Because the variable x is in the exponent position, we will have to become familiar with a new operation, namely, taking the log of both sides. The study of logarithmic functions will occur in 8.R.3.

The opposite (technically the "inverse") of exponentials are logarithms, so to isolate the variable x in the exponent position requires us to take the log of both sides of the equation. To solve most exponential equations requires a four step process:

1. Isolate the exponential variable.
2. Take the logarithm (log) of both sides
3. Solve for the variable
4. Use a calculator to solve

For now, the most useful log rule states that exponents inside a log become multipliers in front of the log. Let's look at example:

Example	Explanation
Solve $5^x = 7$ to the nearest thousandth	Because x is an exponent, we need to use logs.
$\log 5^x = \log 7$	Take the log of both sides
$x \log 5 = \log 7$	The log rule states that the exponent becomes a multiplier – move the exponent to the front of the log.
$\dfrac{x \log 5}{\log 5} = \dfrac{\log 7}{\log 5}$	Now, we can isolate the variable x, by dividing both sides by $\log 5$.
$x = \dfrac{\log 7}{\log 5}$	The $\log 5$ on the left side cancels (reduces to 1), so the variable x is isolated.
$x = \dfrac{0.845}{0.699}$	You will need to locate the log button on your calculator. So, $\log 7 \approx 0.845$ and $\log 5 \approx 0.699$.
$x = \dfrac{0.845}{0.699} \approx 1.209$	To evaluate, $5^{1.209} \approx 7$. We have solved for x when it is in the exponent position.

Name:_____Date:_____

Instructor:_____Section:_____

Guided Examples   Practice

Solve the following for x to the nearest thousandth:	Solve the following for x to the nearest thousandth:
1) $7^x = 60$ This is an exponential equation, so we have to work with logs. First, take the log of both sides so $\log 7^x = \log 60$. Next, the log rule states that the x can move to the front of the log as a multiplier, so $x \log 7 = \log 60$. Next, isolate x by dividing both sides by $\log 7$, so you get $x = \dfrac{\log 60}{\log 7}$. We need a calculator to solve for x. The result is $x = \dfrac{1.778}{0.845} \approx 2.104$. To check, $7^{2.104} \approx 60$.	1) $3^x = 40$
2) $9^x = 100$ This is an exponential equation, so we have to work with logs. First, take the log of both sides so $\log 9^x = \log 100$. Next, the log rule states that the x can move to the front of the log as a multiplier, so $x \log 9 = \log 100$. Next, isolate x by dividing both sides by $\log 9$, so you get $x = \dfrac{\log 100}{\log 9}$. We need a calculator to solve for x. The result is $x = \dfrac{2.000}{0.954} \approx 2.096$. To check, $9^{2.096} \approx 100$.	2) $12^x = 75$

Objective 3 – Evaluate logarithms

In our next undertaking, we will investigate the use of logarithmic scales. Logarithmic scales exist when activities revolve around orders of magnitude. These areas include the magnitude scale for earthquakes, the loudness of sound measured in decibels and the pH scale for acidity.

This will cause us to examine the method to convert logarithms to exponential form. This entails transforming the logarithmic form $y = \log_b x$ if and only if $b^y = x$. Let's look at a challenge:

Evaluate $\log_3 9$. Rewrite the problem as $\log_3 9 = y$. The pattern to transform from log form to exponential form results in $3^y = 9$. There are two methods to find the value for y:

- Use the log form we just discovered in 8.R.2. The result is $\log 3^y = \log 9$. Then the y moves to the front of the log as a multiplier $y \log 3 = \log 9$. Divide both sides by $\log 3$ to isolate the variable y. So
  $y = \dfrac{\log 9}{\log 3}$. Finally, using a calculator, $y = \dfrac{0.954}{0.477} \approx 2$.

- You ask yourself the question, "3 to what value exponent (y) equals 9 ?" $y = 2$. To check, $3^2 = 9$.

Guided Examples	Practice
Evaluate the following logarithms: 1) $\log_5 25$ Rewrite the problem to $\log_5 25 = y$. Then, use the pattern to transform the log form to exponential form, so $5^y = 25$. Finally, ask yourself the question "5 to what value exponent equals 25 ?" $y = 2$. To check $5^2 = 25$. 2) $\log_4 64$ Rewrite the problem to $\log_4 64 = y$. Then, use the log rule to transform the log form to exponential form, so $4^y = 64$. Finally, ask yourself the question "4 to what value exponent equals 64 ?" $y = 3$. To check $4^3 = 64$. 3) $\log_3 81$ Rewrite the problem to $\log_3 81 = y$. Then, use the pattern to transform the log form to exponential form, so $3^y = 81$. Finally, ask yourself the question "3 to what value exponent equals 81 ?" $y = 4$. To check $3^4 = 81$.	Evaluate the following logarithms: 1) $\log_2 32$ 2) $\log_{10} 1000$ 3) $\log_5 125$

Name:_____Date:_____

Instructor:_____Section:_____

Chapter 9 Modeling Our World

Learning Objectives

Objective 1 – Identify and plot points on a coordinated plane
Objective 2 – Graph equations in the rectangular coordinate system
Objective 3 – Decide whether an ordered pair is a solution of a system of linear equations in two   variables
Objective 4 – Find the slope of a line
Objective 5 – Write equivalent exponential and logarithmic equations

Objective 1 – Identify and plot points on a coordinated plane

We will begin to explore the graphic and numeric relationship between values. If these values are related by a function, they are considered variables because the values can change. For now, our focus will be on the graphic representation of how these variables relate.

A function describes how a dependent variable changes with respect to one or more independent variables. The related variables are often written in the form of an ordered pair (independent variable, dependent variable).

Functions are demonstrated graphically using a coordinate plane. The organization of the coordinate plane is as follows:

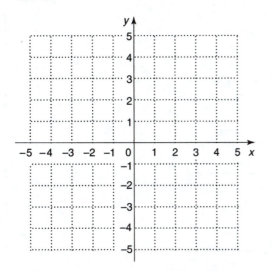

Points (location) on a coordinate plane are described by using an ordered pair as well. The ordered pair is in the form (x, y). To graph a point (identify a location), you start at the origin identified by the ordered pair $(0,0)$. Due to the ordered pair format (x, y), the x coordinate value has you move right (positive) or left (negative) before the y coordinate has you move up (positive) or down (negative).

Guided Examples

Practice

Locate and graph the following points:

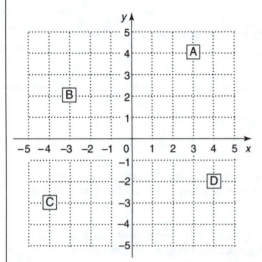

Identify the ordered pair values (x, y) for the given points:

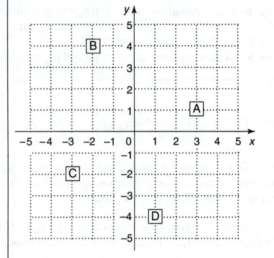

Point A has the ordered pair coordinates $(3, 4)$ because from the origin, you travel $+3$ units (3 units right) along the x axis direction, then $+4$ units (4 units up) in the y axis direction.

Point A: _____

Point B has the ordered pair $(-3, 2)$ because from the origin, you travel -3 units (3 units left) along the x axis direction, then $+2$ units (units up) in the y axis direction.

Point B: _____

Point C has the ordered pair $(-4, -3)$ because from the origin, you travel -4 units (4 units left) along the x axis direction, then -3 units (3 units down) in the y axis direction.

Point C: _____

Point D has the ordered pair $(4, -2)$ because from the origin, you travel $+4$ units (4 units right) along the x axis direction, then -2 units (2 units down) in the y axis direction.

Point D: _____

Objective 2 – Graph equations in the rectangular coordinate system

In the previous section, we plotted points in the rectangular coordinate system. A prevalent and advantageous way of representing functions is to use an equation to combine many points. Equations can be an excellent mathematical instrument when creating, examining and interpreting mathematical models.

A linear function is represented by a straight line, and as such, indicates a constant rate of change. The rate of change does not change throughout the graph of the line and indicates a rate that describes how one quantity changes in relation to another quantity.

To graph linear equations in the rectangular coordinate system, we refer to the linear formula $y = mx + b$. In words, the combined result is:

$$y = mx + b$$

$$\text{total value} = \text{rate of change} \cdot \text{variable} + \text{starting point}$$

If x is the independent variable and y is the dependent variable, then the rate of change can be expressed using the formula:

$$\text{rate of change} = \frac{\text{change in y}}{\text{change in x}}$$

NOTE: For non-vertical lines, the starting point is always on the y axis because it is the value of y when $x = 0$. Vertical lines create an exception. In this case, the starting point may not be on the y axis.

Guided Examples	Practice
Using the following story, put the data into the $y = mx + b$ format, then graph. 1) There is no snow on the ground (starting point $= 0$). A snowstorm is on the way and forecast to snow at a constant rate of 2 inches every 1 hour. So, $y = mx + b$ becomes $y = \dfrac{2}{1}x + 0$	Using the following story, put the data into the $y = mx + b$ format, then graph. 1) Your area is in the middle of a drought. Your local lake is now 5 inches below normal. Several rain storms are on the way and they are expected to increase the lake level by 2 inches every 3 days.

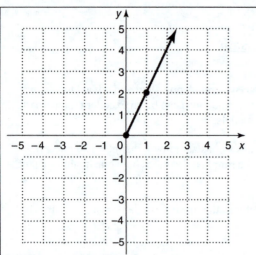

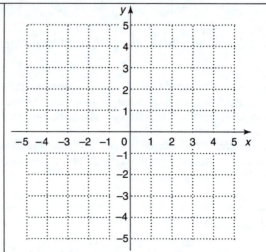

2) Because of recent rain, the local river is already 2 feet above flood stage and a new rain storm is on the way. The water level of the river is expected to rise 1 foot every 3 hours.

So, $y = mx + b$ becomes $y = \dfrac{1}{3}x + 2$

2) The world champion hot dog eater can eat 1 hot dog every 2 seconds. He is starting the contest (start point $= 0$).

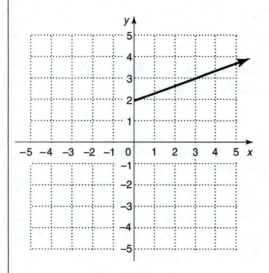

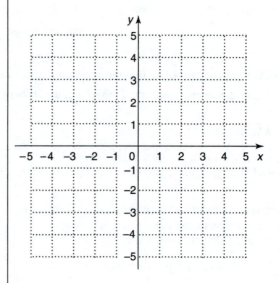

3) You start the holiday season 5 pounds overweight. Your goal is healthy living and a fitness program so you lose 2 pounds every 1 week.

So, $y = mx + b$ becomes $y = \dfrac{-2}{1}x + 5$.

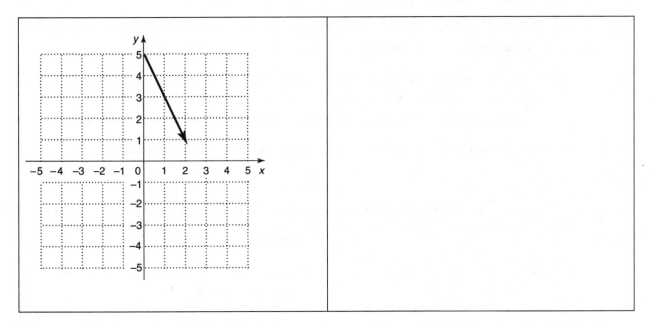

Objective 3 – Decide whether an ordered pair is a solution of a system of linear equations in two variables

Our algebraic challenge in this section deals with a solution that is provided to us in the form of an ordered pair (x, y). We need to decide if the information provided is accurate. Are you a skeptical person? Do you believe everything you are told or do you challenge items you read, see or hear? This is the decision making challenge that you are presented with in this section.

This involves taking the x value of the provided solution and substituting it for x in the specified equation, then taking the y value of the provided solution and substituting it for y in the specified equation. Complete the required algebraic operations then make your determination.

If the left side of the equation equals the right side of the equation, then the equation is balanced and it is determined that the values of the provided ordered pair (x, y) are a solution to the equation specified. If the left side does not equal the right side, then the equation is not balanced and it is determined that the values of the provided ordered pair (x, y) are not a solution to the equation specified. Let's look at two examples:

EX1: Is the ordered pair $(3, 7)$ a solution to the equation $2x + 5y = 41$?

The ordered pair $(3, 7)$ is in the format (x, y), so the x value is 3 and the y value is 7. You substitute these values into the specified equation $2x + 5y = 41$. After the substitution, the equation becomes $2 \cdot 3 + 5 \cdot 7 = 41$. Perform the algebraic operations to support your decision. The result is $6 + 35 = 41$, and finally $41 = 41$.

Does the left side balance with the right side? If yes, you can conclude that the ordered pair $(3, 7)$ is a solution to the equation $2x + 5y = 41$.

EX2: You are planning the itinerary for a Spring Break road trip with your friends. You are given TripTix by a travel agency. The agency calculates you can average 60 miles per hour and your destination is 1,210 miles away. The TripTix plan states that you can make the trip 18 hours. Do you believe their TripTix plan? Is their math correct?

You use your advanced algebra skills to determine the ordered pair $(18,1210)$ and the equation $y = 60x + 0$ (the 0 is because you are starting the trip from your home).

The ordered pair $(18,1210)$ is in the format (x, y), so the x value is 18 and the y value is 1210. You substitute these values into the specified equation $y = 60x + 0$. After the substitution, the equation becomes $1210 = 60 \cdot 18 + 0$. Perform the algebraic operations to support your decision. The result is $1210 \neq 1,080$.

Since the left side does not equal the right side, you can conclude that the TripTix information is not accurate.

Guided Examples

Practice

Apply algebraic operations to the given ordered pair and decide if the data given is a solution to the specified linear equation:	Apply algebraic operations to the given ordered pair and decide if the data given is a solution to the specified linear equation:
1) Is the ordered pair $(1,8)$ a solution to the linear equation $9x - 2y = 7$?	1) Is the ordered pair $(3,10)$ a solution to the linear equation $10x - 2y = 10$?
The ordered pair $(1,8)$ is in the format (x, y), so the x value is 1 and the y value is 8. You substitute these values into the specified equation $9x - 2y = 7$. After substituting the values, the equation becomes $9 \cdot 1 - 2 \cdot 8 = 7$. Perform the algebraic operations to support your decision. The result is $9 - 16 = 7$, and finally $-7 \neq 7$.	
The left side does not balance with the right side. No, the ordered pair $(1,8)$ is not a solution to the equation $9x - 2y = 7$.	2) You find a good deal on a car. The down payment is $\$2,000$ with monthly payments of $\$230$ for two years. The car dealer tells you that you need to budget $\$7,000$ in order to pay for this deal. Is the dealer providing you with accurate information?
2) There are 3 inches of snow on the ground and a big snowstorm is on the way. The storm is expected to dump 2 inches of snow per hour for 5 hours in the area. An emergency is declared if there are 13 inches or more of snow on the ground. If the snow storm is as bad as they predict, should an emergency be declared?	Use the ordered pair $(24,7000)$ and the equation $y = 230x + 2,000$ to aid in your decision.
To describe this situation, use the ordered pair $(5,13)$ and the equation $y = 2x + 3$.	
The ordered pair $(5,13)$ is in the format (x, y), so the x value is 5 and the y value is 13. You substitute these values into the specified equation $y = 2x + 3$. After the substitution, the equation becomes $13 = 2 \cdot 5 + 3$. Perform the algebraic operations to support your decision. The result is $13 = 13$.	
Since the left side equals the right side, you can conclude that an emergency should be declared for the area.	

Objective 4 – Find the slope of a line

To conclude our exploration of linear equations on the rectangular coordinate plane, we focus on the slope of the line created by the linear equation. The slope of a line is the rate of change of the line between two points.

Geography is all about slope – do you travel uphill or downhill to get to campus? Slope is everywhere in architecture design – I have noticed that the slope of the roof of a home in New England is steep while the slope of the roof of a home in the South tends to be flat. Why is that?

There are many ways to think about the slope between two points. To combine these forms, slope is the rise of the run, the change in y divided by the change in x or the incline/decline of a line drawn between the two points.

The coordinates of the two points are specified using the ordered pair format. The first point is written using the general form (x_1, y_1) and the second point is written as (x_2, y_2). The general formula for slope is

$$slope = \frac{rise}{run} = \frac{\text{change in y (up or down)}}{\text{change in x (left or right)}} = \frac{y_2 - y_1}{x_2 - x_1}$$

EX1: What is the slope of the line segment connecting the points $(2,14)$ and $(5,20)$? First, you should place the coordinates for the points in their general form. So the first point $(2,14)$ is (x_1, y_1) and the second point $(5,20)$ is (x_2, y_2). Substitute into the slope equation to get $slope = \dfrac{y_2 - y_1}{x_2 - x_1} = \dfrac{20-14}{5-2} = \dfrac{6}{3} = 2$.

When you are asked to calculate slope, it makes no difference which point you call the first one and which point you call the second one. Since you are working with the same points, the slope of the line is the same regardless of which is the first point and which is the second point. To continue our example, let's determine the first point $(5,20)$ is (x_1, y_1) and the second point $(2,14)$ is (x_2, y_2). Substitute into the slope equation to get

$slope = \dfrac{y_2 - y_1}{x_2 - x_1} = \dfrac{14-20}{2-5} = \dfrac{-6}{-3} = 2$.

Name:_____Date:_____

Instructor:_____Section:_____

Guided Examples

Practice

Calculate the slope of the line segment connecting the coordinates of the two given points:

1) $(3,8)$ and $(5,9)$

First, you should place the coordinates for the points in their general form. The first point $(3,8)$ is (x_1, y_1) and the second point $(5,9)$ is (x_2, y_2). Substitute into the slope equation to get $slope = \dfrac{y_2 - y_1}{x_2 - x_1} = \dfrac{9-8}{5-3} = \dfrac{1}{2}$.

2) $(8,14)$ and $(11,12)$

First, you should place the coordinates for the points in their general form. So the first point $(8,14)$ is (x_1, y_1) and the second point $(11,12)$ is (x_2, y_2). Substitute into the slope equation to get $slope = \dfrac{y_2 - y_1}{x_2 - x_1} = \dfrac{12-14}{11-8} = \dfrac{-2}{3}$.

3) $(-5,-9)$ and $(8,-15)$

First, you should place the coordinates for the points in their general form. So the first point $(-5,-9)$ is (x_1, y_1) and the second point $(8,-15)$ is (x_2, y_2). Substitute into the slope equation to get
$slope = \dfrac{y_2 - y_1}{x_2 - x_1} = \dfrac{-15-(-9)}{8-(-5)} = \dfrac{-15+9}{8+5} = \dfrac{-6}{13}$.

Calculate the slope of the line segment connecting the coordinates of the two given points:

1) $(6,15)$ and $(8,20)$

2) $(5,9)$ and $(8,7)$

Objective 5 – *Write equivalent exponential and logarithmic equations*

The goal of writing equivalent exponential and logarithmic equations is to solve for x when x is an exponent. This causes us to examine the method to convert logarithms to exponential form. The basic properties of logarithms lead us to explore two algebraic procedures. For now, our focus is on common logarithms (base 10). This can be written as $\log_{10} a^x = \log a^x$ and is identified on a calculator as the "log" button.

Our first general rule states that $\log_{10} a^x = x \log_{10} a$. The important portion of this rule is that the variable x can be moved from the exponent and placed in front of the term as a multiplier. For example,

Solve for x in the equation $2^x = 30$. Because the variable x is in the exponent place, we know to use logs to solve this problem. Take the log of both sides, so the result is $\log 2^x = \log 30$. Because of the rule, move the x to the front of the term, so $x \log 2 = \log 30$. To solve for x, divide both sides by $\log 2$. You get

$$x = \frac{\log 30}{\log 2} \approx \frac{1.47712}{0.30103} \approx 4.9069 .$$

The next general rule states requires us to recall from previous notes that $\log_b x = y$ if and only if $b^y = x$. Going forward, it applies that $y = 10^x$ and $y = \log_{10} x$ are inverses of each other.

Solve for x in the equation $4 \log_{10} = 20$. Because the variable x is in the logarithmic expression, the first step is to isolate $\log_{10} x$. This is done by dividing both sides by the factor 4. The result is $\log_{10} x = 5$. Next, we apply the general rule and get $10^5 = x$.

Guided Examples	Practice
Solve for x in the equation:	Solve for x in the equation:
1) $4^x = 18$	1) $7^x = 53$
Take the log of both sides, so the result is $\log 4^x = \log 18$ Because of the rule, move the x to the front of the term, so $x \log 4 = \log 18$. To solve for x, divide both sides by $\log 4$. You get $$x = \frac{\log 18}{\log 4} \approx \frac{1.25527}{0.60206} \approx 2.0850 .$$	

2) $3^x = 35$

Take the log of both sides, so the result is $\log 3^x = \log 35$ Because of the rule, move the x to the front of the term, so $x \log 3 = \log 35$. To solve for x, divide both sides by $\log 3$. You get

$$x = \frac{\log 35}{\log 3} \approx \frac{1.54407}{0.47712} \approx 3.2362.$$

Solve for x in the equation:

3) $5 \log_{10} x = 15$.

The first step is to isolate the $\log_{10} x$. This is done by dividing both sides by the factor 5. The result is $\log_{10} x = 3$. After applying the general rule, $10^3 = x$ or $x = 1000$.

4) $2 \log_{10} x = 5$.

The first step is to isolate the $\log_{10} x$. This is done by dividing both sides by the factor 2. The result is $\log_{10} x = \frac{5}{2}$. After applying the general rule, the result is $10^{\frac{5}{2}} = x$ or $x = 10^{2.5}$.

2) $6^x = 220$

Solve for x in the equation:

3) $2 \log_{10} x = 4$

4) $5 \log_{10} x = 18$

Name:_____Date:_____

Instructor:_____Section:_____

Chapter 10 Modeling with Geometry

Learning Objectives

Objective 1 – Find the perimeter of a polygon or the circumference of a circle
Objective 2 – Find the area of a polygon or circle
Objective 3 – Find the volume of the geometric solid
Objective 4 – Find the missing sides of similar triangles
Objective 5 – Find the unknown side of a right triangle using the Pythagorean Theorem

Objective 1 – Find the perimeter of a polygon or the circumference of a circle

DISCLAIMER: Please note that shapes and measurements in this chapter may not be to scale.

The perimeter of any straight sided polygon is the total distance around the outside of the shape. The perimeter can be calculated by adding together the length of each side. This is true regardless of the type of polygon - a regular or an irregular shape. Perimeter is measured in one dimension, so the answer is in *units* like inches, feet, centimeters or meters.

For example, calculate the perimeter of the following irregular shape.

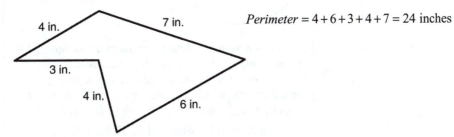

Perimeter $= 4 + 6 + 3 + 4 + 7 = 24$ inches

The perimeter of a circle is called the circumference. The formula for circumference is $C = \pi d$, where C is the circumference, π (*pi*)has an approximate value of 3.14 and d is the diameter (the distance across the circle with the path of the line going through the midpoint of the circle). To summarize, you can calculate the circumference of a circle by multiplying the diameter by π.

For example, calculate the circumference (distance around the circle) of the following circle.

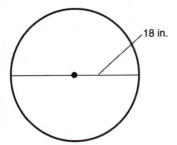

18 in.

Using the formula $C = \pi d$ and plugging in information from the diagram, $C = 3.14 \cdot 18 \approx 56.52$ inches.

Guided Examples	Practice

Calculate the perimeter or circumference from the shapes given:

1)

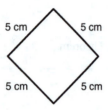

5 cm   5 cm

5 cm   5 cm

Perimeter $= 5+5+5+5 = 20$ cm

2) Diameter = 9cm

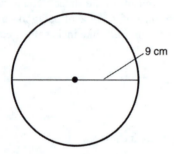

9 cm

$C = \pi d \approx 3.14 \cdot 9 \approx 28.26 cm$

Calculate the perimeter or circumference from the shapes given:

1)

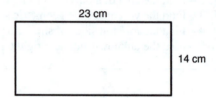

23 cm

14 cm

2)

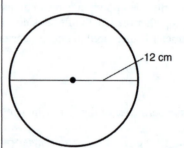

12 cm

3) You and your family are moving to a new home. Of course, you have to have the biggest bedroom. You have not seen the home. Based on the blueprint measurements, which bedroom will you choose if bed room #1 has dimensions of $12,11,9,15$ feet and bedroom #2 has dimensions of $12,12,10,10$ feet ?

Objective 2 – Find the area of a polygon or circle

The area of a rectangle calculates the number of square units inside the shape. For comparison, think of perimeter as the length of fence needed to enclose your yard, where the area of the shape measures the area of the yard inside the fence.

With regards to units, perimeter is one dimensional (from Learning Objective 10.R.2) measured in *units* and area is measured in square units such as square inches, square feet, square centimeters or square meters and is written as $units^2$.

The formula for calculating the area of a rectangle or parallelogram is $A = b \cdot h$, where A is the area, b is the measure of the base of the shape and h is the height of the shape.

As we recall, the distance around a circle is called its circumference. The area of a circle measures the number of square units inside the circle.

The formula for calculating the area of a circle is $A = \pi r^2$, where A is the area, π is approximately 3.14 and r is the radius. The radius is measured as the distance from the center of the circle to any point on the circle.

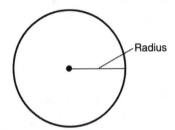
Radius

Guided Examples

Calculate the area of the following shapes:	Calculate the area of the following shapes:
1) 6 in. [rectangle] 35 in. The area of the rectangle is $A = b \cdot h = 35 \cdot 6 = 210$ inches^2	1) [rectangle] 21 cm 6 cm

Practice

2)

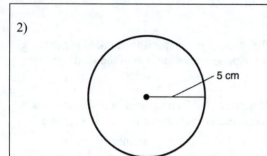

5 cm

The area of the circle is $A = \pi r^2 = 3.14 \cdot 5^2 = 78.50$ cm^2

2)

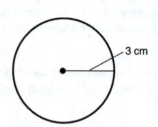

3 cm

3) The student parking lot on your campus needs to be coated. The parking lot is a rectangle shape and measures 400 feet by 220 feet. How much parking lot surface (in square feet) needs to be covered with blacktop?

Objective 3 – Find the volume of the geometric solid

A quick examination recalls that perimeter focuses on one dimension measurement, area speaks to two dimensional measurement, and now, volume is the geometry of three dimensional measurement. This is due to an emphasis on three dimensions – width (w), length (l) and height (h), resulting in cubic units such as cubic inches, cubic feet, cubic centimeters and cubic meters.

In three dimensions, these geometric shapes consist of containers –boxes, food storage and cylinders. For us, our attention will emphasize volume. Specifically, how much material (water, sand, food) a container may hold. We concentrate on two specific shapes – the rectangular prism and the cylinder.

A rectangular prism comprises the same properties but takes different forms – from a shoebox to a cereal box to a pizza box and even a smartphone. The general form is:

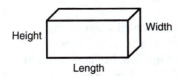

The volume formula for a rectangular prism is $V = l \cdot w \cdot h$ with $units^3$.

A cylinder comprises similar components – from a Pringle's can to canned drinks to some oatmeal containers. The emphasis is on the elements $\pi \approx 3.14$, radius (r) and height (h). The general form is:

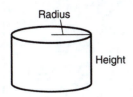

The volume formula for a cylinder is $V = \pi r^2 h$.

Guided Examples   Practice

Calculate the volume for the given geometric solids:
1)

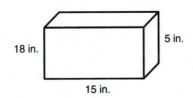

For this rectangular prism, we use the formula $V = l \cdot w \cdot h$. So, $V = 15 \cdot 5 \cdot 8 = 600 inches^3$.

Calculate the volume for the given geometric solids:
1)

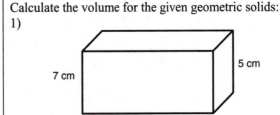

2)

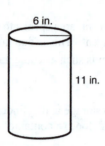

6 in.

11 in.

For this cylinder, we use the formula $V = \pi r^2 h$.

So, $V \approx 3.14 \cdot 6^2 \cdot 11 = 1243.44 inches^3$.

3) You are shopping for candles and notice two types of candles. The rectangular prism shaped candle measures 7 inches by 4 inches by 10 inches. The cylinder shaped candle has a radius of 3 inches and is 12 inches tall. Which candle shape has the greatest volume and the most wax?

Rectangular prism: $V = 7 \cdot 4 \cdot 10 = 280 inches^3$

Cylinder: $V \approx 3.14 \cdot 3^2 \cdot 12 = 339.12 inches^3$

The cylinder shaped candle has the greatest volume and contains more wax.

2)

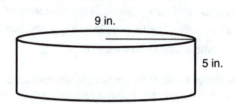

9 in.

5 in.

3) You are thirsty and are presented with two choices at the store. A juice box measures 5 cm by 4 cm by 6 cm. The same juice comes in a can that has a radius of 3 cm and is 6 inches tall. Which shape contains has the greatest volume and contains more juice?

Name:_____Date:_____

Instructor:_____Section:_____

Objective 4 – Find the missing sides of similar triangles

In this section, we explore and discover relationships between similar triangles. We first need to know what similar triangles are. Two triangles are similar if all their angles are the same measure and their corresponding sides are in the same ratio. This can be reworded to mean the same triangle shape but different size.

We need to recognize and identify the corresponding sides of similar triangles, and then write ratios to compare and identify any missing lengths of sides of similar triangles. Let's look at an example of the two similar triangles:

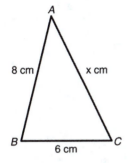 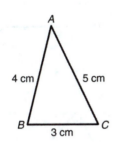

The example states these are similar triangles. As a general guideline, these triangles fit the similar triangle classification – the triangles have the same shape but a different size. Therefore, the corresponding sizes are similar.

We can identify the corresponding sides of these two similar triangles and express the relationship in a proportion or a set of three equal ratios $\dfrac{sideAB}{sideDE} = \dfrac{sideBC}{sideEF} = \dfrac{sideAC}{sideDF}$.

We apply algebraic reasoning to write ratios to form a proportion to solve for the length of the missing side. By substituting the given lengths, we generate $\dfrac{8}{4} = \dfrac{6}{3} = \dfrac{x}{5}$. Lastly, we use the cross product rule to solve for the missing value. We can use either complete proportion to solve for the missing value. To solve, $\dfrac{6}{3} = \dfrac{x}{5}$, then $6 \cdot 5 = 3 \cdot x$, and $30 = 3x$, then divide both sides by 3 to get $10 = x$. The missing length of $sideAC$ is 10 cm.

Guided Examples

Practice

Find the missing side of the similar triangles:

1)

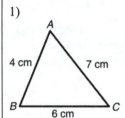

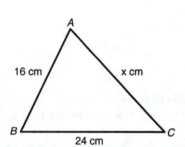

Find the missing side of the similar triangles:

1)

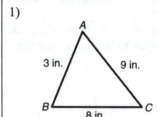

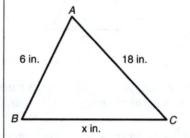

$\dfrac{side AB}{side DE} = \dfrac{side BC}{side EF} = \dfrac{side AC}{side DF}$, substitute $\dfrac{4}{16} = \dfrac{6}{24} = \dfrac{7}{x}$.

To solve, $\dfrac{4}{16} = \dfrac{7}{x}$, then $4x = 16 \cdot 7$, and $4x = 112$, then divide both sides by 4 to get $x = 28$. The missing length of $side DF$ is 28 cm.

Objective 5 – Find the unknown side of a right triangle, using the Pythagorean Theorem

Our challenge in this section is to comprehend, apply and calculate the dimensions of a right triangle using the Pythagorean Theorem. To use the Pythagorean Theorem accurately, we need to become familiar with the 90 degree angle (the right angle), the legs and the hypotenuse of the triangle.

With a right triangle, one of the angle measures exactly 90 degrees. The side that is opposite of the 90 degree angle is called the hypotenuse. This side is always the longest side of the right triangle. The other sides (always shorter than the hypotenuse) are called legs. The right triangle components look like:

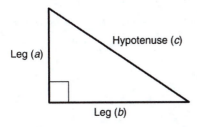

The Pythagorean Theorem states that if you add the squares of the lengths of the two legs of a right triangle, it equals the square of the length of the hypotenuse. Specifically, let a and b be the lengths of the two legs of a right triangle and let c be the length of the hypotenuse. Then,

$$a^2 + b^2 = c^2$$

Guided Examples

Practice

Use the Pythagorean Theorem to calculate the length of the hypotenuse:	Use the Pythagorean Theorem to calculate the length of the hypotenuse:
1) In a right triangle, side $a = 9$ inches and side $b = 12$ inches. What is the length of the hypotenuse? 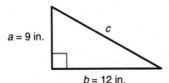 Since $a^2 + b^2 = c^2$, we substitute the values into the appropriate part of the equation. So, $9^2 + 12^2 = c^2$. To solve for c, $81 + 144 = c^2$, then $225 = c^2$. To solve for c, we need to take the square root of both sides. This looks like $\sqrt{225} = \sqrt{c^2}$. Finally, you may need a calculator to take the square root of 225 resulting in $15 = c$. By our calculation, the length of the hypotenuse is 15 inches.	1) In a right triangle, side $a = 18$ inches and side $b = 24$ inches. What is the length of the hypotenuse?

Chapter 12 Mathematics and Politics

Learning Objectives
Objective 1 – Evaluate an algebraic expression in table form

Objective 1 – Evaluate an algebraic expression in table form

In this chapter, you will discover that different methods of counting votes can lead to different results and further investigate the "fairest" method of apportioning Congressional representation among voters. To do this, we will practice evaluating an algebraic expression in table form. You will be given several values and a rule and then the value must be placed into the rule or process to determine the numerical outcome.

For example, fill in the table given the following information:

Value (x) is the number of hours you study	The process $5x + 60$ describes a relationship between the number of hours you study and the score on your next math test (out of 100).
0	$5 \cdot 0 + 60 = 60$
2	$5 \cdot 2 + 60 = 70$
4	$5 \cdot 4 + 60 = 80$
6	$5 \cdot 6 + 60 = 90$

Guided Examples

Practice

Fill in the table given the following information:

1) The growth chart in a pediatrician's office may be modeled after the process formula $2x+7$, where x is the number of months. Fill in the table to calculate the weight (in pounds) of a newborn at various months of age.

Value in months	Process $2x+7$ (in pounds)
1	$2 \cdot 1 + 7 = 9$
3	$2 \cdot 3 + 7 = 13$
6	$2 \cdot 6 + 7 = 19$
9	$2 \cdot 9 + 7 = 25$

Fill in the table given the following information:

1) Since 2001, the process formula $0.4x+11$ describes the poverty rate in the United States, where x is the number of years after 2000. Fill in the table to calculate the poverty rate (in percent) for the given years:

Years since 2000	Process $0.4x+11$ (in percent)
1	
2	
3	
4	

Name:_____Date:_____

Instructor:_____Section:_____

Answers

Chapter 1

Objective 1   1) True 2) True 3) False 4) False

Objective 2   1) False 2) $6+4 \neq 10$ 3) True 4) False 5) True 6) True 7) False 8) True

Objective 3   1) rational, real 2) natural, whole, integer, rational, real 3) False 4) True

Chapter 2

Objective 1   1) $\dfrac{10}{21}$ 2) $\dfrac{6}{5}$ 3) $\dfrac{10}{7}$ 4) $\dfrac{20}{3}$

Objective 2   1) $3.28\cdot10^5$ 2) $1.4629\cdot10^4$ 3) 3.410^{-3} 4) $5.8\cdot10^{-6}$ 5) 6,820 6) 710,000 7) 0.000911
8) 0.02185

Chapter 3

Objective 1   1) $\dfrac{3}{11}$ 2) $\dfrac{1}{5}$ 3) $\dfrac{4}{14}=\dfrac{4\div2}{14\div2}=\dfrac{2}{7}$

Objective 2   1) $\dfrac{83}{100}$ 2) 0.14 3) 57% 4) 60% 5) $75 6) $90 7) 2 tons

Objective 3   1) 0.0738 2) 0.75 3) Yes 4) 25% 5) 5% 6) Yes

Objective 4   1) 3.15×10^5 2) 1.7×10^{13} 3) 3.7×10^{-3} 4) 1×10^{-6} 5) 2900 6) 149,600,000 7) 0.043
8) 0.000006 9) 8×10^8 10) 9.6×10^{10} 11) 2×10^9 12) 1.84×10^3

Objective 5   1) 430 2) 530,000 3) 9800 4) 8.214 5) 439.3 6) 45.34

Chapter 4

Objective 1   1) 100 2) 167 3) $260 4) $4,820 5) 18 6) 25 7) −11 8) −60 9) −10 10) 5
11) 10 12) 10 13) −6

Objective 2   1) $x=16$ 2) $x=9$ 3) $x=6$ 4) 30 students

Objective 3   1) 2 2) 0.061 3) 0.08 4) 7.9% 5) 410% 6)12.42 7) 285 8) $500\cdot4\%=\$20$

Objective 4   1) $x=6$ 2) $x=6$ 3) $x=9$

Objective 5   1) 12 2) 4 3) 4.90 4) $1.29=129\%$

Chapter 5

Objective 1   1) Illinois 2) Kentucky 3) 9,900

Chapter 6

Objective 1   1a) 4 years old   1b) 4 years old   1c) 2 years old

Objective 2   1) 3   2) 8   3) approximately 6.93

Objective 3   1) $x = 60$ 2) $x = 30\%$ 3) $x = 125$

Objective 4   1) 0.058   2) 0.14   3) 0.071

Chapter 7

Objective 1   1) 72   2) 240 outfits   3) $\dfrac{25}{36}$ 4) $\dfrac{5 \cdot 4 \cdot 3}{10 \cdot 9 \cdot 8} = \dfrac{60}{720} = \dfrac{1}{12}$

Objective 2   1) 6.4   2) 11.6   3) $E(v) = \left(2 \cdot \dfrac{9}{10}\right) + \left(10 \cdot \dfrac{1}{10}\right) = 1.8 + 1.0 = \$2.80$

Objective 3   1) $(2)^4 \cdot (7)^3$ 2) $(9)^3 \cdot (4)^2$ 3) 24   4) 120   5) 114   6) 30   7 10,000

Chapter 8

Objective 1   1) linear growth   2) exponential decay   3) linear decay   4) exponential growth

Objective 2   1) $x = \dfrac{1.602}{0.477} \approx 3.358$ 2) $x = \dfrac{1.875}{1.079} \approx 1.737$

Objective 3   1) $y = 5$ 2) $y = 3$ 3) $y = 3$

Chapter 9

Objective 1   Point A: $(3,1)$ Point B: $(-2,4)$ Point C: $(-3,2)$ Point D: $(1,-4)$

Objective 2   1) $y = \dfrac{2}{3}x - 5$

2) $y = \dfrac{1}{2}x + 0$

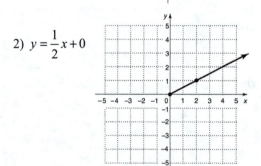

Objective 3   1) $10 = 10$ yes   2) $\$7,000 \neq \$7,520$ no

Objective 4   1) $slope = \dfrac{20-15}{8-6} = \dfrac{5}{2}$ 2) $slope = \dfrac{7-9}{8-5} = \dfrac{-2}{3}$

Objective 5   1) $x = \dfrac{\log 53}{\log 7} \approx \dfrac{1.72428}{0.84509} \approx 2.0404$ 2) $x = \dfrac{\log 220}{\log 6} \approx \dfrac{2.34242}{0.77815} \approx 3.0102$ 3) $x = 10^2$ 4) $x = 10^{3.6}$

Chapter 10

Objective 1   1) $Perimeter = 23 + 14 + 23 + 14 = 74cm$ 2) $C = \pi d \approx 3.14 \cdot 12 = 37.68cm$ 3) Bedroom #1

Objective 2   1) $A = 21 \cdot 6 = 126cm^2$ 2) $A = \pi r^2 \approx 3.14 \cdot 3^2 = 28.26cm^2$ 3) $88,000\,ft^2$

Objective 3   1) $V = 7 \cdot 5 \cdot 13 = 455cm^3$ 2) $V \approx 3.14 \cdot 9^2 \cdot 5 = 1271.70cm^3$

3) Prism: $V \approx 5 \cdot 4 \cdot 6 = 120cm^3$, Cylinder: $V \approx 3.14 \cdot 3^2 \cdot 6 = 169.56cm^3$ CYLINDER

Objective 4   1) $x = 16$ in.

Objective 5   1) $c = 30$ inches

Chapter 12

Objective 1   1)

Years since 2000	Process $0.4x + 11$ (in percent)
1	$0.4 \cdot 1 + 11 = 11.4\%$
2	$0.4 \cdot 2 + 11 = 11.8\%$
3	$0.4 \cdot 3 + 11 = 12.2\%$
4	$0.4 \cdot 4 + 11 = 12.6\%$